The Secret
Language of Life

The Secret Language of Life

How Animals and Plants Feel and Communicate

BRIAN J. FORD

Fromm International
New York

First Fromm International Edition, 2000

Copyright © 1999 by Brian J. Ford

All rights reserved under International and Pan-American Copyright
Conventions. Published in the United States by Fromm International
Publishing Corporation, New York. First published in Great Britain by
by Little, Brown and Company (UK), in 1999

Library of Congress Cataloging-in-Publication Data is available.

ISBN 0-88064-254-8

Manufactured in the United States of America.

Contents

Acknowledgements

In compiling the information in this book I am indebted to some of the wisest of academic minds. The distinguished author, my great friend Brian Aldiss of Oxford, provided me with insights into the way biology has been handled by science fiction. Among those who read the draft manuscript are Sir Colin Spedding, Chairman of the Farm Animal Welfare Council, Professor Peter Biggs, founder of the Institute for Animal Health, Professor Patrick Bateson of King's College, Cambridge, and Professor Pedro J. Aphalo of the Forest Research Institute, Finland. Some have pointed out new relationships between ideas or warned of the risks of infelicities, and all have drawn my attention to points I'd otherwise have missed.

To the many experimental ethologists whose work has guided my thoughts I express undying admiration. Professor Tony Hawkins of the Marine Laboratory at Aberdeen has helped my understanding of the language of fish, and Professor Axel Michelson of Odense University, Denmark, has illuminated the complex subject of insect communication at close range. My descriptions of the diversity of communication in the insect world have been further illuminated by Professor D.

Brian Lewis, whose work at the Guildhall University, London, has pioneered important areas of research, and my understanding of bats and their use of ultrasound owes much to Professor David Pye of Queen Mary College, London. One of my innovations as Zoological Secretary of the Linnean Society of London was an annual full-day conference at the British Association for the Advancement of Science, and David organised the meeting which brought these themes together. It was a memorable conference, and the audience discovered how much we still have to learn about the languages of animals.

I am especially grateful to many good friends, particularly Professor Ifor Bowen and Professor Michael Claridge of Cardiff University; Dame Miriam Rothschild, a perpetual fund of wisdom and experience; Professor Alessandro Minnelli of the University of Padua and Professor Jack Cohen of Warwick University, both great companions who relish the discussion of adventurous ideas. I am grateful to the Committee for the History of Biology, which I chair in London, particularly to Dr Bernard Thomason who organises our lunch-time discussion meetings. Thanks are due to Professor Terry Mansfield for discussions on plant senses and the role of stomata, and to Sir Cyril Clarke, whose research extends from butterfly genetics to the rhesus factor. Among the biologists who have shown generous hospitality in the USA are Professor John Cairns Jr of Virginia State University, a wise and innovative friend, and Professor Fred Addicott of the University of California at Davis, the world authority on abscission in plants. I am indebted to Professor E. O. Wilson of Harvard, with whom I met up in London, for his pioneering insights into sociobiology. My heartfelt gratitude goes to Professor John Corliss, whose writings have inspired a generation of protistologists, to Professor E. C. Hill at Cardiff and the late Dr A. G. Lowndes at Peterborough, whose early teachings proved to be so influential. Among those who facilitated my bibliographical research are the library staff at Cardiff University, and at the Linnean Society of London, the librarians at the Cambridge Periodicals Library and Mr P. K. Fox and his staff at Cambridge University.

Brian J. Ford
Cambridge, 1998

Preface

In this book we will encounter a wonderful world of intelligent animals and clever plants. We are going on a journey of exploration of the wealth of senses and feelings in all forms of life. It is often said that we humans are the only species with emotions, the only ones who think. I wish to persuade you of a fascinating alternative view. The world is actually teeming with sentient creatures, which often possess senses which seem miraculous to us. We are not the only species which communicates; every form of life shares a universal language. Living organisms are talking, listening, detecting information and passing messages to one another. Plants are on the move, animals are busily solving their individual problems, while microbes can look at each other with tiny eyes. Some microbes construct homes out of stone, or snare their prey with traps.

As we shall see, the fashionable concept of humans as superior beings who are uniquely able to communicate must be set in its proper context. The condescending view of lowly and mechanistic forms of life, living through automated reflexes in a world they cannot experience, can no longer stand scrutiny. Our reductionist attitude to science searches for single small causes of specific effects. Science for the new millennium needs to fit these findings together, and to find large, holistic relationships between data and events. Throughout the families of living organisms there are forms of universal communication which transcend evolutionary complexity. If we see a human in distress, or one who is roused to aggression, then we expect to recognise the signals because they are those we experience in our everyday lives. Charles Darwin realised that dogs and apes use recognisable signs of fear and anger in signalling to each other, but we can also recognise distress in non-human species: cats, worms and even in garden plants. The aggression recognisable in dogs is equally detectable in snakes, fish and beetles. You can watch tiny aquatic organisms gathering themselves to spring upon their prey, clearly lurking with intent. This is no anthropomorphism; what we have to do is recognise the extent to which we humans manifest the behavioural patterns common to so many ancient forms of life. They were doing it long before we existed, so these comparisons can hardly be dismissed as imposing human sensibilities on early life forms. They were doing it first.

Our vision of reality derives from the interplay of the human senses. Many other forms of sense in the living world are not possessed by humankind, and other organisms can detect things in ways that we cannot imagine. Some can sense the earth's magnetism, watch the changes of polarised light, see heat rays in the dark, or create an image out of ultrasound. Sea creatures construct detailed pictures of their surroundings through pressure waves in the oceans, or find their hidden prey through electrical currents in the water. Insects do daily problem-solving. Plants turn in the dark, ready to face the morning sun. Tribes of creatures great and small are bonded by their emotions, and use languages of subtle complexity.

Whether they show 'intelligence' depends on your use of the term, and indeed we have little enough knowledge of what this word connotes in human society. There are plenty of people of high IQ who cannot conduct themselves in a thoughtful manner, and *idiots savants* who can draw or play the piano brilliantly, while being mentally impaired in other ways, show how little we know about how our minds work. A semantic debate about the meaning of intelligence is not a part of this argument; I am taking the broadest view of sensory function, problem-solving and communication. The animal world offers many challenges we would find it hard to match. If you want to pit your wits against another species, try building a nest of twigs like a crow, creating a woven shelter like a bower-bird, or damming a stream with sticks like a beaver. A thecate amoeba can construct a delicate flask-shaped shell for itself, using tiny fragments of sand at the bottom of a pond. No human can do anything as intricate. If dexterity, adaptability, the use of environmental features to improve one's environment and problem-solving in unique situations are concomitants of 'intelligence', then let us stop to admire the microbial communities which teem in stagnant water and revel in the abilities of other living organisms to make the best of their surroundings.

Modern human society has become self-obsessed. We see humans with high intelligence, magnificent brains and finely attuned senses; this heady sense of superiority has prevented us from appreciating the great wealth of senses in other species. Brute beasts are seen as mere automatons, and we are currently convinced that we can comprehend them best by looking into their genetics and the chemistry of their cells. Our actions, we think, are finely judged and carefully considered, whereas animals act like chemical factories, governed by blind reaction. They are driven by the need to ensure the survival of their genes above all else. Our visions of the world around us have to be personal and subjective; something you might regard as pleasurable, someone else might consider unpleasant.

To discuss the implications of this world-view, philosophers coined the term solipsism (from the Latin *solus* – 'alone' and *ipse* – 'self'). It is the notion that one can know the world only in terms of personal experience – an intriguing idea, and something of a truism, for each consciousness can know only its own world-view. 'This reasoning implies that we can know only our own notions of what we call other spirits, thus leading, by a *reductio ad absurdum*, to [the] Egoism of Solipsism,' wrote A. E. Taylor in 1894, when the term was first introduced. It may have seemed like a reduction to absurdity then; but it subsequently spawned a huge philosophical literature.

One example is the idea that, if a tree falls and there is nobody there to hear it, then the tree made no sound. There is a similar view that, since the universe is detected by human senses, without them the universe does not exist. These banal ideas have been created by philosophers to rationalise their slight hold on perception. Since our awareness comes to us through our senses, and this awareness is the root of our construction of reality, it is seductively simple to assume that a lack of awareness implies an absence of reality.

My own philosophical view is different. Reality is not contingent upon human cognition. A scientific journal recently published a commentator's view that I am 'an extreme realist'. Perhaps so. I am aware of the physical consequences of a tree falling in the forest, which creates disturbances as it falls that can be scientifically assessed. I am aware of the existence of the physical universe and its marvellously complex interactions, which we can infer through our senses – but which will exist as ever it did long after we are gone, just as it existed before humans appeared. The notion that reality depends on our ability to perceive it shows the extremes of human self-obsession, but it also reveals how far we have moved from a balanced understanding of our place in the affairs of life. Even if a lack of human sensation did imply reality failure, the argument still falls.

What has been overlooked is the richness of sensation in other forms of life. The disturbance made by the falling tree in the forest is detected by innumerable other sensate organisms. Even the patterns of the universe are known by many creatures. Some species use star maps to guide their nocturnal activities. Those time-honoured arguments do not merely reveal a tiresome obsession with self, but show how poorly we appreciate the finely tuned senses of the other organisms who share the planet. Here we will take a tour through living organisms of many types, and we will see how their senses give them information about the world and how they communicate what they know.

The refined nature of human intellect has always blinded us to the marvellous abilities of other forms of life. Make no mistake: molecular biologists have made tremendous strides in unravelling the chemistry of the living cell. However, that gives no indication of how the cells interact to give us the majesty of life on a global scale. It is as if a Martian were to explore the components of an Earthling's television set. Our alien investigator would construct vast and complex lists of interconnected components, but she would learn nothing of how you watch a show, what the programmes mean, or the current commissioning policy of the network – and those are the issues that matter. Molecular biology is like that: it teaches much of metabolism, and the way cells regulate their chemistry, but it tells us little of life in the round. In a new millennium I want us to move towards a holistic view of *organismal biology*. Science has to make the shift from its technical obsession with peering into reductionist crevices, into tinier components of a great and glorious reality. Our modern age is drowning in data. We have more than enough information. The task ahead is to fit our disparate scraps of knowledge into greater syntheses, and learn how living organisms interact. That must be the future of the biological sciences.

There's an anecdote which shows how detached from real biology the molecular biologists have become. A molecular biologist was camping with a friend. They decided to barbecue a sheep, cooked in the old way, to recapture their cultural roots. Up a lane he found a shepherd who said he liked to bet.

'I'll bet that I can count your animals in five seconds flat,' said the scientist. 'If I lose, I'll pay you twenty pounds; but if I win, I claim a sheep.'

Said the farmer, 'That's impossible. It took me an hour. You're on.'

The scientist ran his eye over the flock. 'We have our ways,' he said. 'I work with figures every day.' Then – after the shortest of pauses – he announced his result. 'Two hundred and forty-seven animals in that field.'

'That's incredible,' said the farmer. 'Exactly right. Go and make your choice.'

Within a minute, the scientist was back at the gate.

'Before you go,' said the farmer, 'can I have a go, to see if I can win back the prize? I reckon I could guess your job.'

The scientist chuckled. 'Well,' he said, 'that's probably harder than what I did, so go ahead.'

The farmer beamed. 'I believe you are . . . a molecular biologist,' he said.

The scientist was astonished. 'How ever did you guess that?' he asked.

Said the smiling farmer, 'Oh, we have our ways. Now – give me back my dog.'

Modern biology loses sight of life, for many modern life scientists have little understanding of whole living organisms. Their techniques have retreated into sub-cellular chemistry, losing contact with the wholeness of living and the interactions within communities. The current emphasis on studying the biochemistry of living cells is pervasive, but topics of scientific interest rise and fall like discs in the music charts. These subjects say something about scientific reality as currently construed, but they also reflect the cultural attitudes of the age in which they flourish.

Theories in science reflect preoccupations in society. Moral behaviour was the *Zeitgeist* of the Victorian chattering classes, so moral causes were found for diseases. Every aspect of life was then given a moral dimension. Sexual wantonness was disapproved of in society, so sexual excess was held to be responsible for everything from sheep scrapie to human dementia. When bacteria later became fashionable, germs were likewise blamed for everything. People's teeth were removed wholesale in case bacteria living in them caused some far-off disease in the body – teeth were pulled out to treat arthritis, for instance.

Once the flaws were seen, science turned its back on these bacterial theories. That was itself unfortunate. We now know that most gastric ulcers, long believed to be caused by stress or diet (fashionable preoccupations since the mid-twentieth century), are actually caused by bacteria. The swing in fashion blinded science to that possibility. Most sufferers from ulcers swallow antacid preparations, which make a vast profit for the companies that produce them. It is largely a waste. The bacterium that causes ulcers is *Helicobacter pylori*, and a cocktail of antibiotics can offer a permanent cure. I wouldn't be too surprised if it were discovered that other diseases – irritable bowel syndrome, arthritis, strokes or heart attacks, perhaps even some cancers – are causally related to (unfashionable) bacteria. As this page was being written, cells of *Chlamydia pneumoniae* were identified in the fatty deposits within blocked blood vessels of heart-attack victims. The organism was found in 80 per cent of these patients in a survey at St George's Hospital, London. This timely revelation offers the hope that we may be able to vaccinate against this species in childhood, and thus reduce the incidence of fatal heart attacks in adult life. Fashions in science can prevent us from making new discoveries and from solving old problems.

We are in an era dominated by computers, artificial intelligence, molecular biology and microelectronics. These are all mathematical, precise, reductionist disciplines. Little wonder that we find the nature of life now being defined in terms rich in resonances of these current concerns. The remoteness of such science from biology is highly problematical. In a recently published thesis, a simple robot is described as having the brain-power of a slug. Before long, the argument goes, robots will rival insects, and then humans. This naïve reasoning reveals a failure to understand the complexities of life. The slug, dismissed as 'primitive' in this comparison, knows when to feed and drink, obtaining its own energy supplies throughout its life; it knows how to recognise and avoid unpleasant stimuli, and how to select a sexual partner and undergo sexual display. It can see and it can sense a range of stimuli unknown to humans. The robot needs all the ingenuity of the human mind to make it, and a technical power source to run; the slug can assemble itself out of the debris in your back yard. No electronic device can begin to compare with the astonishing complexity of the 'lowly' slug. Each cell within its body – indeed the components within each cell – reveals more complexity than any robot.

Molecular cell biology and genetics are such marvellously exciting new fields because they reveal much about the basic mechanisms of life. However, they tell you nothing about living. It is like trying to understand the culture of Sumo wrestling by looking at the biochemistry of the biceps. Genetics says nothing about how organisms spend their lives. You can look inside a computer with a magnifying glass, but it won't teach you how to play Doom Warriors. I am more concerned with the ways in which we will think in the future, when we will be able to fit our findings into major new systems of understanding.

In the twenty-first century we will need to change our thinking about animals and plants dramatically. Our present-day attitudes are rooted in the past. Many of the ways in which we interact with animals hark back to our earliest antecedents. The splendour of the hunt, with the huntsmen in their brilliant pink, is a direct connection with the primitive hunters from whom we are all descended. The dissonance between our prehistoric traditions and the growing sense of compassion in the modern world remains a matter of contention. The cultures of mainland Europe have a dismissive attitude to animal welfare, compared with values in contemporary Britain. The intensive raising of calves for veal or force-fed geese for their livers continues apace, and song-birds are often used as items of food. This seems abhorrent to modern Britain, where we witnessed the unusual spectacle of a private member's bill against hunting with dogs being given all-party support when it came to the House of Commons.

Although the bill did not become law, it clearly points to the future. However, hunting with hounds can be sensibly selective. In the English Lake District, where the hilly contour prevents huntsmen from riding fast on horseback, beagles are used to trail and kill a rogue fox which is destroying young lambs. The dogs pick up the scent of the intruder, and run this individual animal to earth. The fox is bound to be killed, and what's more it is only the culprit which the hounds follow. Using guns or gas to kill foxes leaves many of them dying over a prolonged period of time, and there is no selectivity about the foxes that are caught. Against this must be set the essential inhumanity of wanton cruelty to wild animals, and the issues need to be resolved in a manner that will fit us for a more compassionate future.

Of course, primitive humans hunted for their very survival, under different conditions, and according to very different conventions. Stag-hunting with hounds has existed for only four centuries, and research at Cambridge University in 1997 claimed that a stag is unfitted to the chase. Levels of stress hormone were reported to have reached unexpectedly high levels, the exertion being so great that tissue damage was widespread. Other traditions have rarely considered origins. The exotic ritual of the bullfight in Spain, where humans face a weakened and wounded bull in a vast stone arena, is today's version of the ancient contests played out in the Roman amphitheatre. So many of our predecessors' rituals, about which we read with such interest at school, remain with us at the turn of the millennium.

The hunting of the fox and the display of the matador may be minority interests, but the majority of the population is more intimately connected with the exploitation of animals. The mass production of meat has led to intensive farming of an unmistakably inhumane nature. Animals which are clearly sensitive creatures are crowded into concrete pens where they can do little more than survive. The trend towards high profit at any cost has led to a reduction in food safety. New diseases, like 'mad cow disease' (BSE), have taken a massive toll of farm stock and are being related to the deaths of young Britons from a new human spongiform encephalopathy. We have been told to call it a 'new strain' of Creutzfeldt-Jakob disease, but it is very different from that well-documented syndrome and poses insidious new threats (elsewhere I have argued that it seems more closely related to kuru, a disease of the cannibal tribes of Papua New Guinea).

A new strain of *Escherichia coli*, O157.H7, causes a potentially fatal bloody dysentery. Poultry in modern egg farms have been shown to spread *Salmonella* and *Campylobacter*. Proper farm hygiene would act against the spread of these diseases; instead, our exploitive attitude towards animals is already rebounding against the consumer. These diseases, coupled with a growing distaste for animal abuse voiced by the young in particular, pose threats to the existence of the meat industry. A reformed attitude towards our animal associates is a matter of business sense as much as humanity. As the new millennium unfolds, we shall need to abandon such attitudes of exploitation. A more sensitive interaction with these other sentient species must be our aim. I believe that we are surrounded, not by brute beasts we can misuse at will, but by The Secret Language of Life we should appreciate and whose needs we should understand.

For most people, the familiar creatures are domestic animals – cats and dogs, for instance. The relationship between pets and their owners is a constant source of mystery (for those who like pets) or irritation (for those who abhor them). One reason why people like animals is the unquestioning loyalty they sense from their pets, and the fact that pets fulfil the role of friends who – as George Eliot, the pseudonym of Mary Ann Evans (1819–80) emphasised – 'pass no criticisms'. Aldous Huxley (1894–1963) was more trenchant: 'To his dog, every man is Napoleon,' he wrote. 'Hence the constant popularity of dogs.'

Recent research at the University of Central Lancashire by Professor John Archer puts a very different interpretation on our relationship with pets. He suggests that pets act in the endearing way they do effectively to exploit their owners, tricking them into lavishing care and attention that may be at the expense of their duties to family members. Dr Archer summarises his findings by categorising pet ownership as a form of 'social parasitism'. He adds: 'Arguments that pets fulfil a genuine need for affection are misplaced.' This conclusion can be disputed. Many lonely individuals rely on their pets for emotional support; indeed, the loss of a pet can prove to be such a traumatic experience for the owner that psychological counselling is now being proposed. In any event, the emotional bond between pet and owner does not seem to have escaped Dr Archer himself. His own pet is a thirteen-year-old cat named Tabitha. How does his family feel about her? 'We get quite stupid about the cat,' he confesses. So, even a sceptic reveals an unscientific soft spot for a pet. There is a subtle form of language understood by pets and their owners, and pets are clearly sensitive to many of the emotional states experienced by humans.

We are always warned against falling into the trap of anthropomorphism. This is the tendency to find human attributes in animals. You might say that a beetle on its back *despairs* of getting up again, that a trained dog is *dutiful*, that a cat is *determined to have its own way*. These are seductive snares, for they fit our own preconceptions and allow us to graft human sentiments on to animals. Movies like *Babe*, in which a talking piglet airs the feelings of a human, masquerading as those of a farm animal, are seductive. But they are perverse, for they divert us from the realities of the senses animals have possessed since way before the first humans evolved.

Anthropomorphism is not the only trap that awaits the unwary. *Anthropocentrism* is the even more insidious view that humans are so perfectly special that they alone have senses and no animal or plant can have feelings. This is the greatest foolishness of all. No-one will seek to argue about the special quality of intellect which we alone possess, though some might quarrel with the use to which we put it. But senses? There's nothing unique about them. Feelings? Humans are certainly not the only species to share those. When a fond owner says of her dog: 'He understands everything I say,' she's wrong: he doesn't. But he does have an idea of the mood she's in and many of her current intentions. What is true is that *she* understands many of the things her dog says. Dogs, in their own canine fashion, do *enjoy* experiences, they do show *love* for their young, just as they can evince *loyalty* and their eyes reveal *trust*.

The superior stance adopted by our species, as though we can do no wrong and have a unique hold on reality, is not tenable. While we should not confer human attributes on animals, we must open our minds to the emotions and responses, the senses and feelings that other animals *do* possess. It is far more propitious to see echoes of the animals' reactions in ourselves. We learn little by modelling human behaviour in the minds of dumb creatures, but can learn much if we see how they have been feeling and responding for millions of years. We are relative newcomers. There have been animals showing excitement and frustration, devotion and fear, since long before we appeared on this earth. Deny them human feelings, by all means, but we must allow them the right to have their own.

A belief in the complex sensory behaviour of the 'lower' forms of life, as we dismissively categorise them, has punctuated my scientific work for much of my adult life. Sociobiology has given us a revealing legacy, but let us step back from selfish genes and mechanical models to address how animals and plants themselves make sense of their world. The synthesis in the book is a plea for us to value all life and treasure it while we can. We are surrounded by The Secret Language of Life, all communicating through languages of intricate subtlety. Our own mental supremacy is as clear as it ever was – but any claim to uniqueness as sentient beings is unrealistic. We are a component part of the global community of life, not its rulers.

1

Human Senses

There's a battle going on, a war of minds. The current vogue is held by those who extol the complete supremacy of humans over other forms of life. Humans are superior. Our intellect is magnificent, our senses are acute, our language is unparalleled in the realm of living things, and our personal powers are immense. Humankind is unique in adapting items for use as tools. 'Man is a tool-using animal,' wrote Thomas Carlyle (1795–1881), who felt it was the use of tools that made us humans what we are. We are taught that humans alone amongst all species have a language, and that only humans can modify their environment to suit themselves. We have lived with a sense of the supremacy of our species for thousands of years, believing that humans are here to subjugate and control all other forms of life. René Descartes (1596–1650) wrote that, no matter how it appears, animals are incapable of suffering. In the words of Thomas Peacock (1785–1866), writing in *Headlong Hall,* 'Nothing can be more obvious than that all animals were created solely and exclusively for the use of man.'

It is generally assumed that one of the special features which marks humans out from the lowly brutes is the use of tools.

Darwin himself observed that chimpanzees can use a stone to crack open a seed. Later observers watched them use sticks to extract insects from a crevice. Jane Goodall has painstakingly described how a chimpanzee 'fishes' for termites with a stick. If the stick was bent, the animal would turn it round to use the other end, bite off the damaged end, or pick a new stick and start again. Sometime chimps would pick a leafy stem from a tree and painstakingly strip off the leaflets before using it to winkle out their food. This is an example of an animal not merely using a tool, but actually making one. Sea-otters use stones from the sea-bed to crack open shellfish, and finches use plant spines to probe in crevices of bark for insect larvae to eat. Thrushes and other birds choose stones as anvils, and use them to break open the shells of garden snails which they collect for food. Bower-birds collect brightly coloured objects to decorate the interior of their elaborately constructed bowers. Some fish use beads of water, spat at speed, to bring down insects from the air above. Groups from amoebae to insects use sand and stone to construct a safe home. Assassin bugs of the family Reduviidae use lures to attract their prey; one species employs the shed skin of termites as a bait to attract living termites, which it then kills and eats. All sorts of animals are capable of adapting their environment for a range of purposes, and many use tools.

Against the exploitive attitudes stand the animal welfare campaigners. Many movements which campaign for the humane treatment of animals have sprung up in recent years. The death of Linda McCartney in 1998 was celebrated as a symbol of a life devoted to animal issues; she was proclaimed as the animals' princess in parallel with Diana Spencer, who had been hailed as the 'people's princess'. There could be no better illustration of the growing impetus behind organisations that support animals.

These movements were foreshadowed by the writings of Jeremy Bentham (1748–1832), the philosopher and economist who founded the doctrine of utilitarianism. To Bentham, the crucial question was not whether animals could think, but whether they could suffer. In 1983 two philosophers entered the fray. These writers, Peter Singer and Tom Regan, argued that the status of animals should now be recognised. To misuse animals, wrote Singer, is to disregard their interests; to Regan, it was to deny them their rights. Reverence for plant life has also become more fashionable. There is nothing new in this, for bringing plants like mistletoe and holly into the house during the dark days of winter dates back to when they were subject to worship. Crop plants have long been regarded as fertility objects, and there are traces of that ancient belief in many cultural traditions. There is a statue of the Maya corn god at Copán in Honduras, for example, and woven strands of wheat are still made into corn dollies in rural England. There have been reports that the Prince of Wales 'talks to his plants'. Although the practice sounds rather quaint, there are clear benefits in conversing with plants. Our exhalation of carbon dioxide (CO_2) provides a boost to plant metabolism, and anybody who spends time caring for their plants is more likely to note any problems or diseases before matters become serious.

Our thoughts are at a crossroads. The argument against too great an affinity with animals is easily expressed: animals have no rights, for rights imply duties. This is a convenient argument, but a meaningless one. 'Duties' and 'rights' are unrelated, other than in this semantic construction. The paralysed victim of a motoring accident, the congenitally disabled, deaf-mutes and paraplegics have rights to care and consideration, but are in no position to undertake arduous duties. Animals are employed in medical research when they are sufficiently similar to humans to give useful results. If they are so close to humans, the animal campaigners argue that they should be treated more like humans too.

Selfishness and life

A currently held view is that selfishness lies behind life's interactions. People are kind to others because, in the end, it may benefit themselves; humans and other species care for their offspring in order selfishly to perpetuate their kind. The idea dates back several centuries at least. Thomas Hobbes (1588–1697) famously saw life as 'war of all against all' and described life itself as 'nasty, brutish and short': no hint of cooperation or altruism here. In the 1960s the idea of deterministic selfishness became the core of a study by a London zoologist, William Hamilton, who proposed that the cardinal principle of 'aid given to relatives' was a key to the survival of a species. Hamilton attracted a cult following, for this was a controversial new idea. One of his readers was George Price, who resolved to study genetics in order to overthrow the new idea. Instead of disproving the idea of 'genetic selfishness', his studies substantiated the principle. Price and Hamilton collaborated for a time, but Price became increasingly disturbed. He withdrew from formal academic life, turned to religion, and eventually killed himself in a squalid tenement.

A decade later the concept was popularised by Richard Dawkins of Oxford as the 'selfish gene'. The idea found fertile soil at a time when self-interest came to be the guiding principle of personal advancement, with opportunism as the motivation of society. Selfishness could hardly emerge as a prime mover of society without being dignified in the latest fashionable scientific theory, and the 'selfish gene' was that theory. In it we are portrayed as little more than robots, blindly conforming to inbuilt responses that are controlled by and designed for the benefit of the genes using us as a vector. The theory has been criticised for making humans 'no more than animals'.

My view is quite the converse: that animals are, in many ways, comparable to humankind. They communicate with their fellows and care for each other, showing emotions like rage and fear. Play in animals is perhaps an indicator of the celebration of life in species other than ourselves. Many animals play. Birds will repeatedly slide down an icy slope, or tumble in an air-stream, for no apparent reason other than enjoyment. Young elephants play together for prolonged periods, as do young dogs, cats and foxes. At Churchill, Manitoba, hungry polar bears at the end of their winter fast have been seen to approach husky dogs. On one occasion, instead of attacking and eating the dogs, as often happens, a dog and bear were seen to exchange gestures of mutual acceptance and then play together in the snow. At one stage the bear covered the dog with its massive body; later the dog and bear embraced each other. When the bear lay in the snow to cool off from the exertion, the dog stayed attentively close. This was play, not for any long-term benefit to the genes, nor for any ulterior purpose of long-term selfish intent, but apparently for enjoyment. Young macaque monkeys of the Joshinetsu Plateau in Japan even make snowballs. The youngsters gather snow, much as children do, and roll it into balls along the ground. The adult macaques are above such behaviour, though they have been observed to play with snowballs already made by the young ones. Play is an important part of many animals' routine. Interestingly, researchers have ascribed violence among young humans to a lack of play during childhood. It may be that the socialisation induced through play is a crucial part of mental development. Some parents today do not play with their children as their parents did with them, often leaving the children with computer games or other activities in which they indulge on their own. It may be that this leaves a behavioural void which will cause problems in adult life.

The spectrum of animal life teems with examples of animals having fun, and with displays of altruism, loyalty and self-sacrifice. In many ways, the courage and devotion of creatures often militates *against* the survival of their genes. The 'selfish gene' is a crude cybernetic concept, born of the common culture of microelectronics and computer control, and is belied by the complexity of communication and the zest for life which many animals exhibit. If there is a real hidden purpose of life, it is to propagate the germ-cells. All life is essentially microbial. Humans are immortal, for our germ cells go on from one generation to the next, while the great waddling forms that produce them (that's to say, you and me) are nothing more than expendable fruiting-bodies. Humans are an ovum or sperm's way of conquering dry land. So is a marigold.

The notion of the unity of all life is found in many religious beliefs and was a preoccupation of Carl Gustav Jung (1875–1961). In recent years the idea of the planet as an earth-goddess named Gaia has become trendy. A living network cannot function without communication, however, nor without sensory activity on many levels. Now we can begin to understand how the universe of plants and animals is governed by a network of signals, like a vast natural World Wide Web. Life has had its own Internet for more than a billion years.

Looking for simple concepts (like selfishness) as a single governing principle parallels the reductionism of molecular biology in which theorists seek to locate the source of the grand designs of life in the minutest of events. This methodology is that of a technician, not a scientist. Science seeks to fit technical details into the fabric of understanding. I see science as a holistic enterprise. We have been through the era where people sought to model grand concepts as discrete examples through which a greater truth would be perceived. It is time to bring together the scattered ideas and findings of disparate groups of theorists and researchers and weave these threads into the tapestry of a bigger picture.

Admittedly, there is much to be learned by looking at the

enzymes that cascade within our cells, or the cycles of energy that drive our internal chemistry. We can then begin to see how the machinery works. However, far more is to be understood by looking at life in the round. Few cell biologists ever study living cells under the microscope, yet within the single cell we find the seed of our senses. There are tiny cells that have built-in eyes, with a retina and a lens, just like ours. Even microbes can find their way around, tell friend from foe, and decide when to mate and with whom. This book shows how we humans with our traditional five senses are very far from being the only sentient species. The world is teeming with moving, responsive, communicating organisms – animals *and* plants.

Apart from those five senses, humans have other levels of sensation to which we rarely pay much regard. Science does not even admit they exist. But many people will know the reality of subconscious sensation – as, for example, when we wake up five minutes before the alarm clock is due to buzz. And everyone knows how you can suddenly look up to find someone staring at you, as though sensing their gaze boring into you from across the room. If someone in a group yawns (even in a film, or a television programme) then those who are watching feel an urge to yawn, too, and many of them do. We experience warning signs throughout our lives, and take them for granted, but science has yet to show how they work. It has long been taught that babies exist suspended in a blur of unreality for months on end, but it is becoming clear that they can watch you from their earliest days after release from the womb.

It is becoming apparent that the foetus responds to sounds, and reacts to pain with signs of distress. Recent findings reportedly suggest that a newborn baby can even indicate the recall of music heard in the womb in the early months of development. If this proves to be the case, we have been aborting unborn children at twice the age at which they can remember a tune. A leading article in *New Scientist* on 4 April 1998 ended with the following plea: 'We should be open minded enough to allow doctors who treat or terminate foetuses to play safe and use pain control when they can.' A couple of generations ago doctors were prosecuted and struck off for the merest hint of assisting at an abortion, and now some scientists are claiming that the unborn child can respond to music. The unease we can feel at the pendulum swinging to its other extreme could best be gained by reading that sentence again, substituting 'old people' for 'foetuses'.

Single-celled organisms can certainly respond to their surroundings, and show astonishing abilities when constructing homes for themselves. The foetus shows a surprising level of ability to respond, and it would be foolish to conclude that the zygote (the newly fertilised cell from which the embryo develops) is unable to detect its environment. Sensation is a property of single cells, and an ovum will have its own senses. The possibility that these cells might have an awareness of the fertilisation process seems incredible, but the truth is that we know too little about the senses to decide one way or another.

Even our most familiar senses act in ways we do not fully understand. Should you cut yourself on a hidden piece of glass in the sea, it hurts only after you have noticed it. Until then, even a deep cut can remain silent and undetected. This effect can work in the opposite direction. A small child with grazed knees will grizzle and look around for a comforting parent . . . but, if nobody is there to show attention, the crying stops and the child carries on as though nothing had happened. Animals do this too. In laboratory conditions, a baby monkey will flee to its mother if an unfamiliar mechanical toy marches into view. Once it has gained the reassurance it seeks, it will become bolder and try to find out what this new intruder is really like. As many creatures face problems in their daily real-life experiences, we have to concede the extent to which they are working things out and assessing possible solutions. It is true of a wasp colony with a damaged nest, even of a caddis-fly larva with a broken case. They repair the damage, solving practical problems as they proceed. When an ape plays with toys you can see it studying the object intently, working out what to do with it. These animals are thinking, much as we think.

When René Descartes set science on its path of denying thought in such creatures, he fixed on the notion that the senses are merely an organism's means of responding mechanically to the external world. He devised a simple experiment which shows how our interpretation of reality depends very much upon circumstance. Half-fill three bowls with water at different temperatures: one as hot as you can stand, another with ice-cold water, and the third with water that is tepid. Dip your hands into the very hot and very cold bowls for a minute, and then put both into the luke-warm water. The hand that was in the cold water will feel warm; the hand from the hot water feels cold. At the core of Cartesian philosophy was the broad belief that every manifestation of the physical world, from gravity and magnetism to life and love, could be explained solely by mechanistic science. As for thought, it belonged only to those with the power of speech. The word, Descartes said, is the single and certain sign of the existence of thought. This view came to underpin a mechanistic view of life, heedless of the many examples of people who – through disability – could not speak, but possessed obvious intellect. Modern philosophers still cling to this mechanistic view, and molecular biologists are content to believe they are revealing life's realities as they mechanically decode its chemical components. I am convinced that such a narrow approach is no longer tenable. Molecular biology is a tool, not a science. It is a crucial component of the analytical method, certainly; but it will never disclose to us the way living creatures interact in the glory of a global community.

How unique are the five senses?

Our basic senses are well-developed, but they pale in comparison with the sensory abilities of some other animals. We have a good sense of taste, yet an octopus can sense flavours that are a hundred times weaker than anything we can detect. We have a fine sense of smell, but a dog has an olfactory sense a thousand times finer. Our ability to see in poor light is impressive, yet the dolphin has seven thousand times as many light-sensitive cells in its eye. We are good at detecting slight movements, but a cockroach is a hundred thousand times more sensitive to surface vibrations. We struggle to hear sounds below 50 cycles per second (Herz, or Hz for short). Gigantic creatures like whales and elephants communicate over prodigious distances using infrasound frequencies below 15 Hz. Even octopuses and cuttlefish are now known to sense frequencies below 10 Hz. Although we can sense millions of colours, there are garden flowers in which bees can see riotous patterns of which we can only dream.

Some of our senses are subliminal, and we can transmit and receive subtle messages of great power. The traditional way of testing a new drug failed because of this fact. Doctors assessing the drug would make up a placebo, identical in every external respect to the pill containing the drug but containing an inert material instead. They would hand out the placebos to half the volunteers in a trial, and real pills to the other half. Time after time these tests showed that the new drug was marvellously effective, but the results were not borne out by clinical experience. When the explanation emerged, it revolutionised the conduct of medical trials. The doctors were subconsciously signalling to their patients which pill was which. By some minute difference in posture or facial expression, or perhaps the movement of the hand or a twinkle in the eye, doctors intimated to the patients whether they were being dosed with the placebo or the pill. The patients with real pills were thus somehow reassured, and their subliminally acquired knowledge helped to make them well.

These false recoveries are now known as examples of the placebo effect. As a result, the double-blind controlled trial was developed. The trials are *controlled* because there are as many placebos as there are real pills; they are *double-blind* because they are two steps away from the research worker. The real and dummy pills are identified with a code number by the research team. They are then passed to a third party, who takes the anonymously labelled tablets to the nurse who is to administer them, and the nurse then hands them out. This breaks the sequence of subliminal hints which can prejudice the results.

I find it highly significant that we have to go to such lengths. The greatest lesson here is our ability to sense such subliminal signs. Nobody knows exactly what kind of sense this might be, or how it works. Great lengths are taken to eliminate it, but I would be interested to know how to potentiate it, and how to harness this infinitely subtle sense for our own good. When I was first told about this method, during laboratory work when I was a very young aspiring scientist, it was made very clear that we had to be careful, or the patients might detect which was which and recover even if the pills were clinically valueless. There was a far more intriguing possibility: rather than seeking to extirpate the response, why not harness it? If the placebo effect can really rival the power of a medicine, it would be far better to try to use it to the benefit of humankind. Here is a hidden sense we must try to understand. This sense is a subliminal way of receiving information so potent it can act like a drug.

This book presents many observations that show how plants and animals have their own ways of feeling. Plants respond actively to their surroundings, and many of them move when touched. The species of mimosa known as the sensitive plant is famous for that, but the wild rock-rose and the garden pea show similar responses. We shall meet insects that take decisions and spiders that indulge in problem-solving. Animals communicate in extraordinary ways, and the language of the living world will emerge as almost a common form of expression among very different forms of life. Many animals other than humans teach their young to communicate. For example, a chaffinch (if not taught by its parents) is unable to sing properly. The young of many species need parental input at specific ages if they are to grow up to be normal adults. Does this have implications for today's children, sent to watch television or to play on a computer while the parents are preoccupied with bringing in the money? The need to learn has unexpected consequences. We use a range of gestures and movements to signify basic emotions. Many other creatures do this, too: you can recognise fear or submission in a household pet which knows nothing of the intricacies of our society.

Sometimes the symbolism of a gesture is arbitrarily determined. In the Balkans it has long been traditional to shake the head to signify 'yes' and to nod for 'no'. I once took an English youngster to a clinic in Bulgaria, where she was to be examined by a physician who announced with pride that he could speak English. He bent over the child and asked what the problem was.

'It is my ear,' she said, nervously. The doctor looked at the ear nearest to him, and pulled down the lobe.

'Have you pain?' he asked. There was a nod of assent. The doctor then pressed the tympanic bulla, just beneath the ear. He asked again, and another painful nod came in response. He pressed behind the ear, with growing confidence.

'Have you pain?' came his question for the final time. The child nodded firmly.

'Good,' said the doctor, and he crooked his fingers and thumb together in a triumphant flourish, 'this ear is OK.'

I had to explain that a nod meant 'yes', not 'no'. The ear he considered to be clear was actually inflamed and painful. The words 'yes' and 'no' he knew; the problem lay in the fact that the nod and shake of the head had transposed meanings in his culture. Some aspects of body language are not always as instinctive as we think.

Nature and nurture

We still do not know how much of what we do is innate, and how much is acquired. It is the question of *nature* versus *nurture*. Are we the way we are, because that's the way we are? Or do we acquire our current ways because of what we are taught? Some scientists are currently searching for a gene that makes people violent criminals, and for another that could make people homosexual. There is a vogue, as you might expect in this mechanistic era, for believing that everything we do is determined by our genes with the inevitability of a computer program. I believe there is an answer to this conundrum of nature and nurture. Genotype (the genetic constitution of an organism) and phenotype (its acquired adult state) are inextricably linked, and our experience of the world is a powerful factor in releasing or suppressing our genetic potential. It is not a matter of competition between nature and nurture at all, but of the nurture *of* nature; that, and the nature of the nurture. This view gives full weight to the imponderable complexity of factors that affect us. It concedes our fundamental genetic nature, but recognises that how well we can make use of it (and how we come to fulfil our potential) depends very much on the nurture we receive. The resulting cascade of impulses is of unimaginable complexity, and the range of options we consider in our lives arises from the interplay of these inputs and from weighing their merits, one against the other. The reductionist idea that a single impulse must cause a specific response may apply to reflexes, but we step on dangerous ground when we try to extend this concept to judgement and assessment at a cognitive level. The senses act in concert, offering a series of stimuli from which reality is interpreted, and it is from this complexity that free will is born.

Sense organs

All kinds of animals and plants have senses. Microscopic single-celled forms have tiny organelles within the cell which respond to the outside world, while more highly specialised species have organs developed for this purpose. But before we embark on a discussion of sensory abilities in the living world, our own five senses need to be fully understood. Feelings come to us through the senses, and the organs of sense are fundamentally important. An outside stimulus has an effect on a sense organ which is translated into a nerve impulse. This travels to the brain, where the sense itself is finally interpreted. Many of the sense organs can manage an initial level of interpretation: the retina sorts out straight lines and corners, the ear the high and low frequencies. The brain then integrates these signals into the overall experience of the outside world, and a feeling, or an impression, is the result.

The Greek philosopher Aristotle (384–322 BC) recognised the five senses of hearing, sight, smell, taste and touch. Smell and taste are different aspects of the same olfactory sense. Touch has many subdivisions, including the feeling of pressure, temperature and physical contact. Pain is an extreme version of this sense. Other subdivisions can be recognised, including the sensation of weight, the physical position of the body, and the flexing and position of the various limbs. Organs which essentially detect position are collectively called proprioceptors.

Inside the ear is the organ which detects movement of the body and confers a sense of balance. There are other senses on which we depend for the regulation of our actions, like hunger, thirst and tiredness. The most important human senses are hearing and seeing.

When you think of the ear it is the external portion that comes to mind. This skinny flap is the meatus, and acts as a reflector to gather sound and help aim it towards the earhole. This leads to the auditory canal which is 3 cm (1¼ in) long. The important, functioning parts of the ear are hidden for safety within the temporal bone. The middle ear contains the system that amplifies sound and transfers it to the inner ear, where we find the organs of hearing and balance. At the end of the auditory canal lies the eardrum, a thin layer of skin which vibrates as sound waves strike it. This is the structure which captures the sound moving through the air and transduces it to a mechanical form the body can sense. Within the eardrum is a cavity which extends for about 15 mm (about ½ in) and is joined to the back of the throat by the Eustachian tube, named after the Italian anatomist Bartolomeo Eustachio (1520–74) who first described it. As the human embryo develops, it passes through a fish-like early stage in which gill-slits form between the neck and the throat cavity. As the embryo matures all but one of these slits close and become modified. The Eustachian tube is the single gill-slit which remains to remind us of our aquatic heritage.

The middle-ear cavity contains three small bones, loosely linked together. These are the ossicles: the *malleus* (from the Latin for 'hammer'), the *incus* ('anvil'), and the *stapes* ('stirrup'). The ossicles carry sound waves picked up by the eardrum to the *fenestra ovalis* (the 'oval window'). This little tissue-covered window transmits sound to the cochlea (from the Greek for 'snail shell'), a spiral bony capsule filled with endolymph, a transparent fluid. Running along the spiral tube of the cochlea is a set of microscopic fibrils projecting from little hair cells. This delicate array is the organ of Corti, named after its Italian discoverer Bonaventura Corti (1729–1813). The hair cells are set into vibration when a particular sound frequency resonates within the cochlea, and they transmit their impulses to nerve cells that connect with the auditory centres of the brain. There are 16,000 hair cells within this organ, and they can sense movement as small as the width of a single atom.

Different people can sense different sounds. The lowest rate of vibration we can hear as a continuous sound is about 50 Hz (like the 'mains hum' of an electric circuit). Young children can hear around 20,000 Hz (a high-pitched hissing sound). Our ears are most sensitive to notes between about 400 and 8000 Hz, and within this range we can detect differences of pitch no more than 0.03 per cent of the frequency. Our ears are highly sensitive to changes in loudness. Within the frequency range of 1000 to 3000 Hz a change of one decibel can be detected. The effect of sounds on the ear sometimes changes what we actually hear. If you hear a loud sound, muscles contract to dampen down the response of the inner ear. This also happens when you start to speak, and this helps prevent us being deafened by our own voices. Extremely loud notes in the ear actually generate the sensation of notes that were not present in the original sound. These are the result of overloading the system, and produce a harshness or crackling effect a little like a cheap radio turned up too high. This distortion can produce subjective tones, which interfere with listening.

Loudness also affects the nature of the note you hear. If a pure tone is heard, and the volume greatly increased, the eardrum cannot keep pace with the amplitude of the vibrations and the note is perceived as being lower than it is. The effect can lower the perception of the note by as much as a whole tone.

This effect is only recorded when people listen to pure tones. The complex frequency distributions of mixed sounds, which is what we normally hear, do not elicit this response. Our sense of sound can also be modified by masking. Low sounds may set up harmonics of their own within the ear, and these can interfere with the hearing of higher sounds in the mix. It is the unconscious adaptation to masking which makes one raise the pitch of one's voice when speaking to someone in a crowded and noisy room. The loudness of a sound is measured in decibels, dB. This is a logarithmic scale, so that very small differences low down the scale are exaggerated, while the louder sounds are separated by less space on the scale. Thus, a soft whisper measures about 30 dB, a normal voice about 60 dB, and shouting measures 90 dB. The loudest shout ever recorded measured 119 dB (and there is a record of a scream measuring 128 dB). The blue whale *Balaenoptera musculus* can emit pulses of sound rated at 188 dB. This could cause unconsciousness in a human, and these sounds can be heard over a distance of 800 km (500 miles).

Our sense of balance and equilibrium resides in the vestibule and semicircular canals of the inner ear. The semicircular canals are delicate bony tubes, like tiny jug-handles, arranged at ninety degrees to one other, one in each dimension of movement: up and down, forward and back, left and right. Lining the fluid-filled canals are fine sensory cells, much like the hair cells of the organ of Corti, and inside the canals are fine grains of calcium carbonate (chalk), known by the scientific term otoliths or, colloquially, as ear sand. When your head tilts, the otoliths slide around and stimulate the hairs beneath as they make contact. When you spin round on the spot, it is the motion of the otoliths that causes the sensation of dizziness. The turning sets the fluid in the semicircular canals into movement, and the grains of chalk keep moving after you have stopped. The cells within the semicircular canals respond by signalling the continued movement. It is little wonder that the brain becomes confused.

Hearing is a remarkable sense. We can gain so many impressions from the surrounding world. Characteristic overtones can indicate when a speaker is stressed, and for sightless people the sounds in a room can indicate much of its shape and configuration. Many other organisms can hear better than us and, as we shall see, human hearing is a crude device compared with the sense in many moths, bats and birds.

The human eyeball is a rounded structure 2½ cm (about 1 in) in diameter. It bulges out at the front, where the transparent cornea is found. The eye is a camera, looking at the world. It has the same components as a camera: a lens at the front, an iris to regulate the light and 'instant film' – the retina – at the back. The outer layer is the rubbery protective coating, the sclera. Five-sixths of the surface of the eye is covered by this tough white layer. At the front of the eyeball it merges into the bulging, transparent cornea. Beneath this layer is the choroid, richer in blood-vessels than the sclera, which covers 60 per cent of the eye. The choroid runs into the ciliary body (which supports the lens) and the iris (which opens and closes in response to brightness) at the front. Inside the choroid is the retina, which is sensitive to light.

The cornea is a tough, transparent dome. Its five layers are just slightly yellow-tinged to cut down on the violet and ultra-violet which would otherwise damage the eye. This is the structure, rather than the lens, which focuses the image. The overall focusing power of the human eye is about 60 dioptres (a lens one dioptre in strength has a focal length of one metre, so this is 60 times stronger). Of those 60 dioptres, 40 dioptres are due to the cornea, and only the remaining 20 dioptres are the focusing power of the lens. That is why the shape of the cornea

can be surgically modified to change the focus. The first operation to make these corrections was radical keratectomy (RK),
which involved surgically cutting the cornea to alter its shape.
This is a risky operation and is rarely encountered now.
Occasionally light would 'flare' on the modified cornea, making
it hard to see into the light, and this led to a modest weakening
of the globe of the eye – both potentially troublesome problems. It has been replaced by photo refractive keratectomy
(PRK) in which a laser is used to shape the front of the cornea.
The direction of the burn is controlled by computer software,
which fully compensates for eye movements. The surgeon peels
away the epithelium which covers the front of the cornea, and
then the laser ablation takes up to half a minute. Afterwards,
most patients feel modest discomfort, which can be relieved
with eye-drops, whilst a few have an uncomfortable few days.
Rest is important after this new treatment, and patients are dis-
couraged even from watching television. So far this treatment
has proved to be valuable in cases of astigmatism and also mod-
erate short sight. The structure of the lens is remarkable. It is a
fibrous structure made of many transparent layers and is normally an optically perfect lens-shape. Man-made lenses have
many aberrations which distort the image and cause spurious
colour to be seen, but the human lens and cornea are better corrected than anything we can make.

The photographic aperture of the human eye is $f/2.1$, which
makes it rather slower than a professional camera lens. The
human retina, on the other hand, is far more sensitive than any
man-made roll of film. At the rear of the eye, the light beam
reaches its destination: the retina itself. Here it encounters a serious design fault. Before the light can reach the light-sensitive
cells, it has to pass through a layer of nerve fibres. The retina is
a layered structure, the outer layer being rich in blood-vessels
which nourish and feed the cells. However, retina forms as an

outgrowth of the front of the brain. During its development, as it emerges and folds in the embryo, it forms inside out! Each light-sensitive cell ends in a fine nerve fibre which can carry the impulses away to the brain. Instead of the nerves emerging at the back of the cells, they emerge at the front. This means that the retina is not coated by the light-sensitive cells you would expect, but is covered with blood vessels and the nerve fibres carrying the impulses away. The 130 million light-sensitive cells lie beneath that layer. Not only does this reduce the clarity of the image, it also means that the nerve fibres have to find a way out if they are to get to the brain. The central place where they bunch together and burrow out through the retina naturally cannot contain any light-sensitive cells at all. This is the blind spot. Each of our eyes is completely blind in this small central region of the middle of the retina.

To see just how blind it is, cover your left eye with your hand and then, with your right eye, look at the left-hand spot. Move the book towards your eye, or away from it; about a foot away the right-hand spot vanishes. This is where its image falls on the blind spot.

●   ●

There are two types of light-sensitive cell in the human retina. They are known, from their shape under the microscope, as rods and cones. Over 95 per cent of them are rods, which are sensitive to low light levels and respond only to light intensity, not colour. It is with your rods that you see the landscape on a moonlit night, when they can increase their sensitivity by 75,000 times. You may have imagined that the landscape looks strangely colourless and eerie because of the colour of moonlight. That's not the reason. Moonlight is reflected sunlight, and because the moon is a reasonable reflector, the light that reaches us is not very different from normal white light. Colours should seem as colourful under moonlight as they do under sunlight. The landscape seems monochromatic and hauntingly different because the light is dim. The cones, which detect the colours, cannot work when light levels are low. What you see on a moon-lit night is worth appreciating, for you are seeing the world almost as if you had only rods in the retina. The entire retina has rods, while the cones become more tightly packed towards the centre. For this reason your field of view is broader on a moon-lit night, when the rods take over, than in bright light, when the cones come to the fore.

Colour vision needs more light, and this is where the cones come in. There are only 6 million cones among the 130 million cells of the retina, and the nearer to the centre of the retina, the more densely the cones are packed in. Directly in line with the pupil (alongside the blind spot) is an area where the rods and cones are so tightly packed that they produce a yellow-coloured rounded area, the yellow spot or *macula lutea*. In the very middle lies the *fovea centralis*, which contains nothing but tightly packed cones. This is the area of greatest visual acuity of the eye. In general the eyes of all animals resemble simple cam-eras in that the lens of the eye forms an inverted image of objects in front of it on the sensitive retina, which corresponds to the film in a camera. It is often said that the brain 'inverts the picture' but that is a silly argument. Upside down on the retina *is* the right way up for the brain.

The normal spectrum gives us our sense of vision, though for some people (mostly males) red-green colour blindness means that they are unable to distinguish between some shades of red and green. The fact that red means 'danger' and green means 'go' in the Western world is confusing for millions of men. In some rare instances normal colours confuse the brain. Sufferers from some kinds of autism cannot make sense of what they see, for the normal world breaks up into fragments of random colours. This is also true of dyslexic patients with a condition sometimes known as 'scotopic sensitivity syndrome', in which words on the printed page seem to make little sense. Two Melbourne specialists, Olive Meares and Helen Irlen, have shown that if these people wear tinted lenses, the letters suddenly make sense. The coloured overlay does not have to fill the entire field of view, and patients sometimes feel more comfortable with slightly different tints on different occasions. Some dyslexic children have learned to read by using coloured overlays, and as they mature they learn to read without them. This certainly adds a new dimension to the traditional idea of 'looking at the world through rose-tinted spectacles'.

The healthy eye at rest stares at infinity. Its focusing effect, which allows us to sharpen details closer up, is known as accommodation, though the lens stiffens a little with advancing age and thus becomes less able to accommodate. A young child can see details down to as little as 6 cm (2½ in) from the eye. As the lens ages, the eye loses its ability to fine-tune the focus of close objects. During the twenties, the closest a normal eye can focus increases to about 15 cm (6 in) and is more like 50 cm (20 in) by the age of fifty. From then on, the lens of the eye cannot usually accommodate objects as close as working distance, so they have to be held at arm's length. This condition is presbyopia and it is common in elderly people.

Cones are capable of resolving far finer details than the rods. Rods resolve brightness, but they are wired together in groups. There are about 1000 times as many sensory cells as there are nerve fibres in the mammalian optic nerve, so many of the retinal cells have to share their output. Because of their shared nerve fibres, the rods produce an image of a relatively coarse nature. Cones, on the other hand, are individually connected. The nerve cell from each cone goes directly to the optic nerve, so each cell sends its own discrete information. Since there is no pooling of data for cones, they can offer the greatest resolution. That is why they are concentrated in the very centre of the retina. Our eyesight in the middle of our field of vision is exceedingly good, in terms of both colour and fine detail. To protect the cones from too much brightness there is a brownish pigment in the outer layer of the retina. When there is too much light for comfort, tiny granules of this pigment move to positions around the cones where they protect them by screening out the excess light. They take a moment or two to move into place, during which time bright light can seem painful. Once the cones are properly protected the eyes are said to have become *light-adapted*.

The rods need to be sensitised by means of a visual pigment, visual purple (rhodopsin), that forms inside each cell. The visual purple is bleached when light shines upon it, and it can only be synthesised by the rods when light levels are low. As a result, rods do not function properly in bright sunlight. If you watch the pupil of someone who steps from bright daylight into a gloomy room you will see the pupil open immediately, flooding the retina with enough light to see properly. This response takes a tenth of a second. But the person concerned will find that several seconds pass before they can see again. The opening of the pupil is not enough by itself, for the rods need to synthesise visual purple before they can function. From the moment that the light levels drop this process starts, but it takes a while before there is enough visual purple for the eyes to work. Once the visual pigment has formed and the eyes are able to see in gloomy conditions, the eye is described as *dark-adapted*. The production of visual purple depends on vitamin A, and a lack of this vitamin in the diet naturally leads to night blindness. Under normal conditions, it takes 150 milliseconds after light has entered the eye for the sight to be registered by the brain.

The coordination of the movement of our two eyes so that they point the same way is complex. The nearer the object, the more the axes of the eyes must converge. If you look at the top of this page and maintain your focus on it, you will find that everything else in the room is not only blurred, but doubled. Now look from the book to the opposite wall and you'll see that there are two images of the book, out of focus before your eyes. The change of focus is the easy part to grasp, but the hidden beauty of the system is the fact that the eyes diverge a little as you shift your gaze to a distant object. If the system does not work, double vision results. The extent to which the eyes converge helps us to judge distance.

There is an additional function of the eyes, which we do not yet understand fully. They have a nervous connection to the pineal body in the middle of the brain. This is the source of melatonin, which helps to regulate the sense of time and seems to be connected to the state of our mood. This may be the controlling factor of seasonal affective disorder (SAD) or 'winter blues'. The connection with the pineal is fascinating, for in many creatures the pineal body actually *is* an eye. The most remarkable example must be the middle eye of the lamprey. In this species of fish, the pineal body develops a lens, retina and optic nerve and seems to help the animal detect sources of heat.

Protecting the eye is vital. There are several anatomical adaptations which help to preserve this vital sense organ. Most important of these are the eyelids, which close to seal the eye from the outside. They are lined with a delicate tissue, the conjuctiva, which also covers the visible sclera (the white of the eye). Our eyelids blink shut unconsciously every six or seven seconds, more often if there is wind or dust particles in the air. Any sudden movement towards the open eye also causes an immediate, involuntary blink. Around the edges of the lids are the many small Meibomian glands, named after their discoverer

Heinrich Meibom (1678–1740). The glands secrete an oily liquid which helps to lubricate the eyelashes and the interior of the lids. The eyelashes help to keep particles out of the eyes when we are watching the world through half-closed eyelids. In the corner of each eye are the openings of the lachrymal glands, which produce tears. Life began in the sea, and the 'inner sea' in which our cells live is salty to this day. The composition of tears is very much like that of sea-water. They contain lysozyme, an enzyme which acts as a powerful disinfectant and breaks down microorganisms. Lysozyme was a discovery of Alexander Fleming (1881–1955), who found it occurred in many bodily secretions. It was a crucial discovery for Fleming. This is what primed his attention for broad-spectrum substances which could kill bacteria, and made him ready to understand the significance of the penicillin fungus when first he saw its effects in 1928.

The clouding of the lens known as cataract is a common cause of visual impairment. Senile cataract is the most frequent form of the condition. It occurs in the elderly, and typically affects both eyes at once. The first signs are streaks that radiate across the lens from its centre, or small spots that start to form in the body of the lens. In time they spread and coalesce, impeding vision and eventually causing blindness. During the progress of the disease, the lens shrinks and eventually separates from its capsule. At this stage it is considered suitable for removal. Contact lenses or spectacles help restore normal vision post-operatively. In India I found old village doctors who tell of the way cataracts have been treated for centuries. The doctor would pluck a sharp thorn from an acacia bush. This is needle-sharp, readily available and, having been irradiated by sunlight, it has been sterilised. The point of the thorn was thrust through the edge of the cornea, impaling the lens and forcing it down into the vitreous. A twist of the thorn allowed it to be withdrawn and discarded (an early example of a disposable surgical instrument). Although the sufferers were left without a lens at all, they could at least perceive blurred outlines. This partial restoration of sight pre-dated modern medical science by a thousand years at least.

The eyes of some other creatures can hugely improve on the performance of the human eye, but mostly as a result of adaptations to special circumstances. All eyes are multicellular versions of the simple eye-spot within microbes. There are fifty different types of eye in the animal world. Philosophers argue endlessly about the way they could have developed, and about the selective pressures that have given rise to this great variety of eye types in the animal world. All eyes have a lens, but its nature varies enormously; they all have a retina, but these have developed in surprisingly different ways. All eyes have evolved by separate cells assuming specialised functions within the eye, rather than relying on parts within one cell to do it all. The structure of the eye in other creatures has always been portrayed as giving them a blurred and fragmented view of nature. This is, I am sure, untrue. The fly perching on the edge of your plate is looking back at you. It can certainly see images that are coherent and meaningful, and its eyes suit it for its life just as our eyes suit ours.

The sense of smell is housed at the back of the nose. When we inhale through the nose, air is drawn through undulating, thin turbinate bones within the nasal cavity. The olfactory tissues covering these bones inside the nasal cavities are rich in blood supply and nerve endings, and any active molecules in the air can be absorbed by the liquid layer. The nerves then detect them, recognise their special qualities, and send appropriate signals to the brain. There are seven basic types of sensory nerves in the olfactory epithelium, and there are seven basic scents now recognised which may be related to these different sensors. These are categorised as: camphor, musk, floral, mint, ethereal, pungent and putrid.

The most promising avenue of current research is based on the finding that substances with similar odours have molecules of similar shape. If the shape of a molecule is altered in the laboratory, its odour changes. Although we have categorised seven basic scents, the human nose can detect over 10,000 different odours with its 5 million sensory cells. Curiously, although only three genes are necessary to code for the receptors within the retina, up to 1000 genes are needed to produce the sense of smell. There is a clear difference in the way women and men detect scents. A woman can detect a form of musk at concentrations 1000 times lower than a man. Chemical odours are important signals throughout the living world, and they are collectively known as pheromones. The uterus of the female prairie vole triples in size if she senses male pheromones, and some male moths can scent a female two miles away. Many animals are known to possess a vomeronasal organ which is specialised for the detection of pheromones. Some research now suggests that the same structure may also exist in humans. It is certainly true that concoctions containing human pheromones are commercially available, though I am not aware of any research which proves that they work.

The organs of taste are buried in the tongue, and these taste buds can distinguish four main categories: sweet, sour, salt and bitter. There are additional taste buds in the roof of the mouth and in the upper end of the pharynx. The power of the sensation varies with each category. Sugary flavours can be detected at dilutions down to 1:200, salt at 1:400, sour at 1:150,000 but bitter tastes can be sensed when diluted to 1:2 million. The mucous membrane in these regions is dotted with minute projections (papillae), each of them in turn covered with between two and three hundred taste buds. The papillae at the back of the tongue, the circumvallate papillae, form a V-shape with the point towards the throat and can detect bitter tastes. The tip of the tongue has the taste buds which can sense sweetness, while salt and sour tastes are sensed by the papillae on each side of the tongue. Different people have different amounts of taste receptors. Some have more than 10,000 in total, whilst others have no more than 500. Those with fewer sensors lack the refinement of sweet and salty flavours, though their ability to detect bitterness is unimpaired.

The nerve fibres from each papilla join together to form a nerve cord which travels out through the base of each papilla, and eventually connects to the brain. A substance that is to be tasted has first to dissolve, so insoluble substances cannot ordinarily be tasted. The strength of the taste is related to temperature. Very cold substances tend to deaden the taste buds. This is why cheap drinks are best served very cold, and helps to explain the freezing temperature at which much modern beer is dispensed. The coldness prevents the customer from discovering how inferior the drink really is. Until the popularisation of the refrigerator, drinks could only be served cool, but not very cold; this is why traditional concoctions (like English bitter) are presented at cellar temperature. The brew can be tasted to the full, so care must be taken to ensure that the taste is pleasing. Cheap modern beers need to be drunk near freezing point, which prevents the lack of nuance being detected by the customer.

Many foods are savoured as much by smell as by taste. The full sense of taste owes much to the related sense of smell. People given scrambled egg to taste while blindfolded will not know what it is if the nose is pinched shut. They have to smell it too, for there is not enough of the basic taste in this food to make it recognisable by taste alone. Many animals have a sense of taste thousands of times more acute than ours. Sharks, for example, can taste blood in the water over 1 km (½ mile) from a wounded animal in the sea.

Many related senses are grouped under 'touch', last of the five traditional senses. The basic organ of touch is a sensory body known as the Pacinian corpuscle, found on the fingertips. These sense organs were discovered by Abraham Vater in 1717, but the discovery was largely ignored until 1840, when a description was published by Filippo Pacini (1812–83) of Pisa, and they are known after him to this day. As Pacini showed, the nerve endings in these little sense organs swell out to produce bulbs. There are pain receptors throughout the body, even in the tissues around the intestines. It has to be said that touch is one of the least specialised of the senses, but this is largely because we do not normally cultivate it. Blind people develop their sense of touch to a high degree, and can read the slight impressions of Braille at lightning speed.

The skin is the main tactile area of the body, and is also the largest organ, measuring over 2 sq m (21½ sq ft) in area. The organs of touch are unevenly spread across the body, being more tightly packed on the fingertips than across the back. In the skin lie 200,000 receptors for heat and cold, 50,000 for pressure and tactility, and up to 3 million for pain. The tactile sensors are most tightly packed on the genitals, lips, fingertips and face, but even here they can only distinguish two objects if they are separated by more than 3 mm (⅛ in).

Pain and discomfort are sometimes detected by the organs of touch. In these cases there is a strong psychological component. Painful wounds are often ignored until someone draws attention to them. The sense organs are interrelated, and between them they present to the mind a detailed picture of the outside world. Other organisms inhabit very different worlds, and have their own senses to cope. As we shall see, many of them have finer feelings by far, and have unexpected and extraordinary organs with which to picture their world

2

The Mammalian Mind

Wrestling with the complexities of the human mind is difficult enough, as we have already seen, but it is even harder to fathom how other mammalian species operate. Many mammals carry out complex tasks which astonish us. The way a beaver dams a stream is a remarkable example, for no two streams are alike and the beaver displays considerable ability to assess a situation and work out what to do next. The beaver's actions are not merely task-directed, but are goal-oriented – the animal clearly has a mental picture of what it wishes to attain. The completed dam, with its concealed chamber within the beaver's lodge, defies the simplistic notion of an animal driven by instinct, like a robot.

Mammals experience their own emotions: fear and insecurity, triumph and passion. Zoo animals display neuroses like long-term human prisoners in solitary confinement. Dairy cattle, when deprived of their newborn calves, moan as though in misery for days on end. We are not the only species that dreams, either. Take your dog out to chase rabbits in the fields, and then settle down for a relaxed evening by a log fire. Your trusty hound will whimper and cry out as it sleeps at your feet, legs moving as it hunts, its eyelids betraying the visions it sees, dreaming of the chase – until a gentle nudge from your foot will awaken it just enough to lose the thread. We are taught that dreams are a uniquely human prerogative. That is wrong. Our animal relatives dream, too, and have done since long before we as a species ever existed.

Many animals parallel humans in their mental behaviour. There are cats with nymphomania and anorexic dogs. Clinical hysteria has been observed in goats, cows and even chickens. There have been studies of psychotic cats, impotent bulls, and neuroses in a range of domestic pets. If young male baboons are moved from familiar surroundings into a threatening environment, like a cage with older males and a human audience, they can develop deep distress and even convulsions. Infant macaque monkeys separated from their mothers huddle in groups, inconsolable and miserable. Human societies are accustomed to life's victims – misplaced individuals in an oppressive community who never flower, but remain passive. Studies have shown that this also happens in the animal world. Terriers in a litter often gang up on one of their number, making it a social outcast. The rejected pup will stand around listlessly at the edge of the action, head and tail down, with little interest in food. But what happens if the poor creature is placed in a new group of non-aggressive spaniels? Within half an hour the animal recovers its purpose in life, starts interacting and feeds well. Comparisons with people in an office with unfriendly and inflexible managers, or isolated within an unsympathetic family, are easy to draw.

There are even animal drug addicts. Rats can be experimentally habituated to alcohol and guinea-pigs to marijuana. Wild animals are known to use drugs. Pigeons are particularly fond of over-indulgence in marijuana in the wild, and the koalas' diet of eucalyptus can sometimes send them into a trance. Coffee is said to have been discovered around AD 900, when a young

Abyssinian goat-herd noticed that his flock became frisky and excitable after eating the berries of the coffee tree. The Yemeni tradition of chewing 'qat' (the leaves of *Catha edulis*) is said to have begun when sheep grazing on the leaves became stimulated and active. There are many other mammals that show a similar addictive tendency. Cats adore the catnip plant, and can become 'high' on it. Elephants know when the juice of a favourite plant is fermenting, and will travel a considerable distance to experience intoxication. So humans are not the only users of stimulants. Nor are we the only creature to show bitterness, depression and the intensities of grief. Separation from a loved one can be as marked in the animal world as it is in human society. For example, monkeys separated from their families show huddling, withdrawal, self-mutilation and a refusal to eat. Who is going to dismiss these findings as 'anthropomorphic'? These are the ways of living mammals, and humankind is just one of the tribe.

It is often assumed that animals are somehow constrained by their senses, and that their senses are forced upon them because of their lifestyles. As one commentator has put it, a sloth hanging lazily from a leafy branch does not need to have much mental agility, whereas a dolphin hunting at speed and in cooperation with others needs fast responses and an active brain. This becomes a self-fulfilling prophecy. In my view it is more illuminating to envisage mammals as using their minds in the ways they need. The sloth and the dolphin are both highly evolved mammals and must surely have comparable mental abilities. Dolphins use their brains to catch fish at speed, while sloths use theirs to adapt to an unobtrusive life in a hot and humid jungle. It is their ecological niche, their behavioural destiny, to which their minds are adapted, and not the adoption of a lifestyle which fits the brain.

Mammals have finely developed senses which give them a

remarkable ability to understand their surroundings. Many are territorial. Elephants in Asia currently kill, on average, more than one person a day, almost always because peasant farmers have encroached on their traditional territory. The attacks are fearsome and terrible, yet show the elephant's ability to direct its anger at the human invader. In one example, an enraged elephant confronted a mother and her baby. It lifted the baby from the mother's arms, laid it carefully to one side, and then attacked the adult human it saw as its enemy, trampling her to death until the remains were unrecognisable. Territoriality of this kind shows how animals can 'possess' property. It has long been argued that it takes the formation of a legal framework to define the ownership of property, yet here are mammals showing this very trait. Many species of bird are territorial, too, showing that the recognition of possessed territory is not confined even to the mammals. Salamanders defend their own territory, so the trait is not even restricted to warm-blooded vertebrates. Even insects defend their plots. The possession of a stretch of land is found in animals very different from mammals.

Mammals can show sensitivity and care, altruism and self-sacrifice. They are complex individuals, and we have to be careful before we design experiments to assess their abilities. For instance, it is well known that dogs have an excellent sense of hearing, and can detect notes far beyond the highest frequencies detectable by the human ear. The usual way of testing a dog's hearing is to transmit short bursts of sound and watch how the dog responds. For other animals this simple technique gives misleading results. Guinea-pigs are widely said to have poor hearing because they do not respond to notes played to them in the same way as dogs. Why is this? Whereas dogs jump up and pay attention, guinea-pigs respond to an unexpected noise by freezing in position, and keeping perfectly still. This misled early investigators into believing that guinea-pigs' hearing was deficient. In fact, the problem was with the scientific method-ology, and not with the animals' ears.

Pioneers of sonar

Bats are mentally alert animals. They live in colonies, and their navigational skills in finding their way home are well-developed. They recognise each other, and some even share food: vampire bats, for example, often drink more blood than they need, then regurgitate it for neighbours in their colony who haven't fed lately. Their main interest to us here is that they possess a sense unlike any of ours. Bats can emit high-frequency pulses of ultra-sound and use the echoes to construct a mental image of their surroundings or their possible prey. These navigational sounds are above the range that human ears can detect, but bats also squeak at lower frequencies which fall within the human range. These high-pitched utterances are sometimes heard by children, whereas their parents are unable to detect any sound. Most species of bat send out sonar waves and analyse the reflected sound patterns as they fly. If this seems hard to envisage, think how we interpret reflected signals (albeit of light) when we explore at night by torchlight. A bat's navigational sense is something like an ultrasonic version of torchlight. As early as the eighteenth century, experimenters showed that blind bats could fly, this ability being largely unaffected by their visual handicap. Deaf bats, on the other hand, became uncoordinated in flight. The importance of a bat's ears was known long before the rea-sons became apparent.

There are a few species of bat which catch their prey in their mouth. Most species flick forward the web of skin between their hind legs to snatch an object out of the air like a baseball player catching a ball. The bat then inspects what it has caught and takes it in its mouth if it likes what it sees. If you toss some gravel into the air while bats are hunting at dusk you may see them follow it as it falls, scanning it by sonar and deciding it is not worth catching before they resume hunting. A young bat may be too inexperienced to know the difference, and can be induced to catch one of these little pebbles. It will carry it briefly and make a visual inspection before discarding it and flying on. Next time it recognises the sonar signal of gravel and ignores it.

The ability of bats to detect obstacles in the dark has often been demonstrated by confining them to darkened chambers containing rods placed at irregular intervals. The bats can fly unerringly past the rods, even at high speed. Wires a mere 1 mm thick are rarely touched, and it is not until the thickness is no greater than that of a human hair that bats will fail to negotiate the obstacles. The width must be sufficient to reflect the ultra-sonic pulses emitted by the bats, and if a wire is too thin for the wavelength to be reflected, there is no means by which the bat can detect its presence. These echoes are faint. The sound signal that comes back to the bat may be thousands of times weaker than the original pulse it emitted. For us, it would be like trying to hear a whisper above the cacophony of a crowded indoor swimming-pool. This explains why many bats have such elaborate and enlarged ears, for they help pick up the faint returning echoes. The continuous movement of the bat through its field of ultrasonic pulses gives it an ever-changing view of the solid objects in the air around it. However, not all species of bat use echolocation. The flying foxes or fruit-bats of the tropics fly during daylight and use senses much like those of a conventional mammal: sight, hearing and scent. I have observed fruit-bats in Asia and tropical Australia, and still marvel at their aerodynamic skill and speed of response.

Producing the sound is one thing; detecting and analysing it is a very different task. The sound is received by the paired ears, each carried on the insect's legs, and relayed to two giant cells –

the omega cells – inside the body within the first thoracic ganglion. They process the pitch of the sound and the relative intensity with which it is detected by the ears on the two forelegs. The cricket's brain is able to process this result, and the insects can detect the direction of an incoming signal to within five degrees of the compass.

Elephants can hear sounds which we cannot. Unlike bats, which send out ultrasonic beams as an aid to navigation (and cannot be heard because they are far too high), the elephant uses a form of sound which is too low for us to hear: infrasound. As we have seen, the lowest frequencies humans can hear as a distinct note are about 50–60 Hz, but it is now known that elephants speak to each other at a frequency below about 20 Hz, at which frequency humans hear nothing at all.

The intensity of this infrasonic communication is surprising. At 1 m (3¼ ft) from the animal, the energy can reach 117 dB, the sound of a very loud rock concert. Close inspection of the elephant's forehead shows that it flutters in sympathy with the sound issuing from the great vocal cords within the larynx. To a casual observer, this is the only clue that the animal is vocalising.

The male (bull) elephant has phases of sexual awakening known in India as *musth* (meaning 'intoxication'). A normal elephant wanders around, feeding, eating and grooming, with little sense of dynamism. The bull in musth changes dramatically. The head is held high, the trunk is raised and the gait changes from a slow shambling to a purposive stride across the bush. Bulls in musth will fight, sometimes to the death. The penis oozes urine in a trickle, and the lachrymal glands on the cheek start to dribble their secretion down the animal's face. The term musth is a good one, for the animal is truly intoxicated. It is high on a rush of testosterone, and fears no enemy.

These elephants find their mates through infrasonic communication. When the cow elephant comes into season she is mounted by one male after another, and she calls to her mates across several kilometres of empty space using these powerful sexual calls. At the Pittsburgh Zoo, giant loudspeakers were specially constructed for this investigation. Each speaker weighs 90 kg (200 lb) and has a volume of about a metre (30 cu ft) to reproduce sounds below the range of human hearing. Through them, the mating calls of a sexually aroused female elephant could be relayed across the Namibian bush. Video recordings show how receptive males stop feeding, turn to face the direction of the infrasound and raise their ears. After listening intently for several moments they set off towards the source, stopping every few hundred metres to listen for further messages.

Perhaps the most vivid demonstration of the power of infrasound from the Pittsburgh Zoo field research is in the form of a night-time video taken in infrared. The scene was virtually invisible to human eyes, and no extraneous sound could be heard by human ears. The tape shows a herd of a dozen elephants, some old, some very young, meandering through the scrub. Once or twice a pair of old males near the rear of the group stop as though listening for a distant signal, while the younger animals continue to browse. The youngsters are playfully interacting with older group members, paying no attention to the distant infrasound which the males can clearly hear.

Suddenly the whole group freezes into immobility. None of the elephants moves. After a few seconds, all begin to turn to face the direction from which they have travelled. The movement is purposeful and coordinated. The animals raise their ears and stand together motionless for several minutes, clearly listening to a distant voice. Even the skittish youngsters remain still. The immobility is remarkable, and the tape looks for all the world as though someone has set the freeze-frame control. If the tape is played through fast, there is still no sign of movement in the image. After a couple of minutes of this stillness the elephants lower their ears, turn and march off across the scrub-land. It is an intensely vivid illustration of the power of infrasound to the elephant community.

Deep-sea denizens

Perhaps the most widespread use of sonar in larger mammals is in the marine species, the dolphins, porpoises and whales. Many people will have heard recordings of their haunting cries, which travel long distances under water. The sounds a creature emits are roughly related to its size. The best way for us to hear the rapid and complex call of a little song-bird is to record it, and play it back at a slower speed which we can more easily study. Perhaps we should try something similar with the choruses of whales. If we were to speed up their utterances (which often sound like sequences of staccato sounds) we would reconstitute a more continuous call closer in line with our own range.

Sound is crucially important in the lives of these sea-mammals, much as it is to us. It is not only used for communication: a dolphin can disable its prey with a burst of high-decibel sound. These creatures also utilise ultrasound for direction-finding. The bottle-nosed dolphin can emit pulses of ultrasound at frequencies up to 120 kHz and can hear frequencies above that. Their left and right ears are at different heights on each side of the head, a structural adaptation also found in owls, and this helps them to identify the phase of sounds as a further aid to direction-finding.

Whales are remarkable navigators, and like elephants they can communicate using infrasound to maintain contact during lengthy migrations. The best places for feeding are not the most suitable sites for breeding, and the different sites are sometimes thousands of kilometres apart. Baleen whales feed on the crustacean *Euphasia superba*, krill, which is commonly found in the Arctic waters where they feed on the plankton nourished by up-welling waters of the sub-polar seas. The sea temperatures are too low for newborn whale calves, so they migrate to warmer tropical waters to give birth. The tropical populations of plankton provide the whales with a little food, but it is a hundred times less than they consume in the shoals of krill, so they rely on energy reserves stored as layers of blubber until they are back in the colder waters. The young, meanwhile, are fed on mother's milk until they have themselves built up a layer of blubber sufficient to insulate them on the cold trek north.

Humpback whales, *Megaptera novaeanliae*, make similar lengthy migrations. Adults I have observed feeding off the shores of the north-eastern USA swim south to winter in the Caribbean. The population living in the north-eastern Pacific travels to the Mexican coast for the winter months, while the humpbacks of the Norwegian coastal waters swim past the British Isles to breed during the winter months off the coast of North Africa. Whales possess magnetic organs of sense, which can detect the earth's magnetic field, and they use their highly developed senses to enable them to make these lengthy journeys. The young follow their parents along the routes as they learn the way. With their acutely developed sense of hearing whales are often disoriented by the barrage of submarine sounds produced by human activity. The sounds of ships' engines can destroy the sonar environment on which sea-creatures rely.

The cetaceans (whales, dolphins and porpoises) also use sound and ultrasound to locate their prey. They can emit bursts of high-energy sound in pulses lasting less than a millisecond, and respond to the echoes just like bats. The higher speed of sound in water makes it important to keep pulses of ultra-sound very short. A porpoise can detect an object the size of a large marble (25 mm) at a distance of over 75 metres (roughly

250 ft). To help focus the energy wave sent back as an echo from their prey, cetaceans have a skull shaped like a reflector to focus the echoes. Buried inside the tissues of the forehead is an oval sac filled with oil. This is known as the 'melon', and it acts as an acoustic lens to further concentrate the pulse of energy.

How do they detect sounds, without external ears? The most likely mechanism for their sense of echolocation seems to involve air spaces between thin bones in the skull. Some of these layers of bone can apparently vibrate in sympathy with the sound waves, and this seems to be how echoes are detected. There can be no doubt about the efficiency of this echolocation system, for dolphins are now known to have an ability to detect echoes which is close to the theoretical maximum.

The dolphin as ally

Mother whales and dolphins are devoted to caring for their young. There is much evidence of cooperation, of care for injured individuals in the school, of attentiveness during birth. It is the dolphin, however, that has the reputation for being friendly to humankind. Some workers have even claimed that the dolphin, with a brain much the same size as ours, is of comparable intelligence. It is an absurd idea. Although these marine mammals have well-developed senses, their intelligence is not much different from that of a domestic dog.

Brain size is irrelevant, or a shrew would be a million times less intelligent than an elephant. If there is a single criterion, it would be the ratio of the cortex to the medulla: the relationship of the outer layer of the brain to its overall volume. Humans have an extensive cortex, which is why it is folded into ridges. Dolphins have smoother brains because the ratio of surface area to volume is much the same as in other mammals. They do not have super-high intelligence or mental powers which compare with ours. Yet they can communicate verbally. Their beautiful and haunting sub-aquatic songs, and the excited clicks and chuckles they utter to their human handlers, clearly constitute a more complex language than we imagine.

Dolphins have a lengthy history of association with humans, and legends show how long they have been construed as intelligent. Their associations with humans date back to prehistory. Over 20,000 years ago palaeolithic cave artists at Levano, in the Engadi Islands off the coast of Sicily, painted pictures of dolphins and men together. In Crete the dolphin was worshipped as a god. When the cult spread to mainland Greece, dolphins were respected as humans in reincarnated form. When Apollo was said to have guided ships through storm-tossed seas to the safety of Criss, near Delphio, he did so in the form of a dolphin, and it was believed that Apollo's son Icadius was saved by a dolphin after being shipwrecked. He was borne safely to dry land, and after he was set ashore the seat of Apollo was established nearby, on Mount Parnassus.

A man riding on a dolphin appeared on ancient Greek coinage and on wax seals. One legend tells of a notable musician returning home by sea after winning a competition. He was robbed of his prizes and thrown overboard by mutinous sailors. Dolphins rescued him and carried him safely to port where he was able to report the offence in time for the sailors to be captured and punished. The Greek philosopher Aristotle (384–322 BC) wrote on the friendship which exists between dolphins and people. Three centuries later, the *Historia Naturalis* of Pliny the Elder (AD 23–79) contained an account of a dolphin which regularly transported a boy across the sea to school, and added, 'Ampholicians and Tarentines testify to touching dolphins which were friends of little boys.' He also recounted a tale of a monkey in a shipwreck being rescued by a dolphin and transported safely to shore. A century after that, the poet Oppian described how dolphins helped fishermen near Athens to catch fish from the shore. The fish in a bay were prevented from escaping by dolphins which would drive them towards shallow water, where they could be caught by men with tridents. The dolphins would be given their allotted share by the fishermen. Some of these stories have counterparts in the modern world. It has been reported that Brazilian fishermen still cooperate with dolphins which drive schools of fish towards their nets.

Accounts of dolphins coming to the aid of people are many. During the mid-1970s a dolphin known as Beaky was regularly seen around the Isle of Man. He gave youngsters rides upon his back, trying to lift them clear of the water. Later he became known in the waters around Cornwall, where he was said to have saved several lives. One was of a diver who got into difficulties off Land's End and was borne back to his boat by the dolphin. Several swimmers claim that their lives have been saved by dolphins off the coast of Florida. A dolphin which accompanied ships across the Cook Strait in New Zealand for many years was protected by law, and a Governor's Order in Council was passed to make it an offence to harm the species. Swimming with dolphins is a form of therapy that has been used with success in treating children with mental disabilities. It has enabled autistic children to communicate and in some cases to show improvement beyond all expectation. The participation of the dolphins is itself interesting, for they clearly become involved in the process of therapy and seem to enjoy it. So great are the benefits to the young patients that one centre at Key Largo, Florida, is able to fund referred patients through health insurance.

It seems that dolphins can learn by studying their own kind. In an American marine park, a dolphin was painstakingly trained to move a buoy by pulling it. When this animal became ill and was removed from the tank, another dolphin was selected to perform the task. This dolphin preferred to move the buoy by a different method, pushing it along with its snout. As this worked just as well, the trainers allowed the animal to do this in its performance. When this second dolphin suddenly died, another in the group took over, immediately pushing the buoy along without any training. For two days in the following month this dolphin would not take part in the performance, so a fourth dolphin took over. He immediately performed by pulling the buoy along with his snout, just as the first-trained animal had done. It was concluded that the dolphins could see how to perform by watching the performer, and the memory of the different techniques stayed with them.

Dolphins show many levels of emotion. They pine for a lost mate and show distress if separated. A paper published in the *Journal of Mammology* recounts the depressed behaviour of a young female dolphin which, after having been caught on a hook, was placed in a marine tank to recover. She remained listless and unable to support herself to breathe, so a male was brought in as company. He spent much time in attendance, holding the female with his beak so that she could breathe in comfort, and her demeanour immediately improved. Within weeks she was well on the way to recovery and the two spent much time playing and swimming together. Three months later she became ill again, as a result of an infection of the original wound, and the male showed increasing signs of distress as her condition worsened. At her death, he uttered a loud cry. From then on he would not feed, but circled the tank aimlessly, uttering cries, until he too died three days later. It is certainly clear that dolphins care for each other when sick or disabled, and become distressed when a group member dies.

Many other mammals pine for a lost partner. They are not alone: pairing birds, such as swans, are said to do so too, and so do some fish. That these animals feel the loss intensely is clear from their responses. Emotions are widespread in the animal world. To claim that we are the only species to suffer distress is to deny the evidence of our observations. It also dignifies humanity in a way that is unwarranted: there are many other organisms, and they too have their feelings.

Clever Hans

People have trained dogs and hawks from before the dawn of history, so there is no questioning the ability of these creatures to learn. Telling how they do it, and when they've done it, is not so easy. Clever Hans proves the point. Wilhelm von Osten was a nineteenth-century Prussian aristocrat who believed animals could be trained. In his view, the reason why animals were not as mentally able as humans was that they were not taught properly. He raised a Polish stallion and gave this fine horse regular instruction in mathematics, using skittles and an abacus, and writing answers with chalk on a blackboard. A correct answer was rewarded with a slice of fresh carrot, the horse's favourite food. After two years he announced that the animal was a mathematical prodigy. He named it Kluge Hans ('Clever Hans'), and claimed that it could perform as well as a twelve-year-old child. Tests proved it – the horse could perform mathematics with complete accuracy, tapping out answers with his hoof on the ground. In the face of sceptical disbelief, von Osten petitioned the Emperor Wilhelm II, who agreed to set up a committee of investigation. The committee included a magician, a veterinary specialist with a knowledge of horses, and even a member of parliament. They were unable to find fault: Clever Hans did indeed appear to be a mathematical genius. He would tap out the numbers on a card every time von Osten selected one for the horse to see.

Once a professor arranged for von Osten to hold up many of the cards without first looking at the figure displayed. He also fixed Clever Hans with blinkers, which allowed the horse to see the cards, and to hear von Osten's voice as he announced each number displayed, but prevented the horse from looking at his master. This produced chaotic results. Clever Hans tapped aimlessly with his hoof, not once providing the right answer. The horse must have been picking up subliminal cues from his trainer's bodily and facial movements. If a blank card was held up, Clever Hans could be induced to tap out any number until the experimenter used slight head movements or a slight flaring of the nostrils to stop the horse from tapping further. The horse was reading slight reactions from the onlookers, rather than numbers on the cards. Von Osten was greatly upset by the revelations, and sold the horse to another animal training enthusiast who was still convinced that animals could be educated.

The extent to which animals can be trained was scientifically analysed by Ivan Pavlov (1849–1936) who taught dogs to associate feeding with the ringing of a bell. He showed that a dog would still salivate eagerly when the bell sounded, even if no food was offered. This gave rise to the concept of the conditioned reflex, and seemed to consign animal learning to the category of automatic reflexes which by-passed any form of thought. Although this taught us much about conditioning, it said little for learning. Animals certainly learn. And it is not only dolphins that learn new lessons from their fellows.

Cattle-grids have been used for over a century in South Wales. Made from tubes or strips of steel laid across a pit in a gateway, they allow wheeled traffic, bicycles and pedestrians to cross, but provide an insecure foothold for cloven-hoofed creatures like cattle and sheep. The strips can take their weight without trouble, but the animals sense that their hooves will slip from the smooth metal surface and they won't step onto the grid. This simple device kept roads open for traffic and people, while being a secure barrier to farm animals. That is no longer the case. During the 1970s a sheep, somewhere in the valleys, found a way to cross. Other sheep turned up to watch, and reportedly learned to copy the example. Sheep are now said to be commonly found wandering the streets from which they used to be excluded. These animals have a reputation for stupidity, yet they learned a new trick and taught it to one another. The secret is simple – they lie down as they approach the grid, and roll over. Once on the other side (a single roll is enough to take them across the gap, feet properly positioned to stand up on the far side) they jump up and trot away.

In the spring of 1997 came reports that sheep in the New Forest village of Bramshaw, near Southampton, had found another means to defeat the grids. One sheep lay across the grid, allowing others to scramble over, using the fleece of the volunteer as a foothold. These animals show an ability to observe, and to modify set patterns of behaviour in the light of what they see. They learn slowly, compared with the might of the human mind. However, it does reveal mental processes which amount to problem-solving. It has been too easy for us to assume that this form of thought is confined to ourselves: these examples show that learning and mental adaptation occur at many levels of mammalian life.

Applied animal intelligence

Apart from tool use in the wild, some mammals can be trained to use apparatus in an artificial environment. A classic example is given by a tame white rat. A food supply is provided for this inquisitive creature. The problem is that the food lies on a shelf, well out of reach. By training the rat through successive steps, it can acquire the technique of reaching the food. Between the rat and the food shelf is another platform, to which it can climb using a lightweight ladder set up in position. Up runs the rat, the food now only half as far away as it was, but still out of reach. The rat now notices that the ladder up which it climbed has a string attached. The string passes over the top of the frame, and hangs down where the rat now stands. For a while the rat runs up and down the ladder, inspecting the string. After a few moments of apparent contemplation, it reaches up to the string and starts to pull. Holding it in its teeth, pulling with its fore-paws, the rat gently pulls the string towards itself. As the string is pulled down, the ladder is raised until its lower end rests securely on the edge of the halfway platform. This is what the rat knows it must achieve. It drops the string, runs up the ladder which it has hoisted into position, and grabs the food at the top.

The memorisation of a sequence, and recognition of criteria which show that the sequence is completed, show something of the power of the rat brain. Rats use extraordinary ingenuity in their lives as feral creatures associated with human habitation. They will even attack a human foe if the occasion arises. It is the fact that rats seem to display intelligence that makes them so feared and disliked by the human community.

Sensing disasters

Animals have long been thought to be able to predict earthquakes. In a report of the earthquake at Callao, Peru, in 1745, it was recorded that dogs left the town shortly before the tremors began. Exactly the same phenomenon was recorded before the earthquakes at Galcahuasco in 1855; Owens Valley, USA, in 1872; Assam, India, in 1897; and the Mexican quake of 1907. Immediately prior to the Agadir, Morocco, earthquake of 1960, stray animals (including dogs) were seen leaving the town in large numbers. Much the same happened before the 1963 disaster at Skopje, Yugoslavia, and at Tashkent, capital of Uzbekistan, in 1966.

Changes in animal behaviour were taken by the Chinese in 1975 to presage the magnitude of the 7.3 Haicheng earthquake. They evacuated 90,000 residents only two days before the quake, which destroyed or damaged 90 per cent of the city's buildings. The following year an earthquake in Tangshan, east of Beijing, was heralded at the British Embassy by the agitated barking of a golden retriever owned by the staff, and in the same year at Udine, Italy, the residents were awakened by the incessant barking of dogs prior to the first tremor. Scientists in China, Russia and Japan are looking into behavioural change in animals as a possible means of assisting earthquake prediction. Zoo animals have also been said to change their behaviour before an earthquake. All this does suggest that animals can provide warnings to human populations. Taken in total, their senses are broader than ours, and we should acknowledge the extent to which they are attuned to stimuli that our senses cannot detect.

Can animals foretell the future?

The agitation of animals before an earthquake strikes has been taken to mean that they can see into the future, but that is more than the reports warrant. Before the earth movement of the quake itself there are subtle changes. There may be alterations in local geomagnetic activity, slight foreshocks, or an imperceptible alteration in the angle of the ground or slight local bulging. A sign of this latent activity is given when the level of water in a well unexpectedly alters in response to geological changes deep underground. Seismograph records of the Haicheng earthquake show many slight foreshocks in the period before the earthquake began. It is more realistic to assume that the finely attuned senses of animals can detect these signs, causing them to feel unease which shows itself as changes in behaviour. In my view, we are more likely to be witnessing the effects of animal physiology, rather than a psychic ability to foretell the future.

Many other stories of animal behaviour are testimony to their extraordinary sensitivity rather than to any psychical abilities. An example is how the pet of an elderly person reacts as the devoted owner becomes increasingly ill. If the two live together there is little mystery about the change in the pet's behaviour, but what if they live apart? Take the case where a much-loved dog goes to stay with the family of the owner's daughter. Changes in the pet's behaviour have been taken to show that the animal is somehow tuned into the changing health of the distant owner. Such explanations overlook the more likely reason: the animal may be picking up cues from the mood of the son or daughter following a visit to the ailing parent.

Another example is the case where two dogs, bitch and pup, went for separate walks but ordinarily lived together. One afternoon the youngster was attacked by a mongrel out on its own. Some days later, the young dog's mother encountered the mongrel while being walked by her owner and immediately attacked it. Does this elicit some 'psychic sense' on her part? There are more mundane explanations. Perhaps the young dog communicated the event to the bitch, which was determined to avenge the attack on her progeny. More likely still is the possibility that she detected the reaction to the mongrel of her owner, who responded with some apprehension at the sight of the aggressive animal. Perhaps most plausible of all is the possibility that the odour of the aggressor that had lingered on the pup was recalled by the bitch. We need to look for practical explanations. More exotic phenomena may call for increasingly unlikely explanations, but in so many cases there is a clear and feasible explanation readily to hand. We should not search for metaphysical explanations when science can offer a working hypothesis.

A farmer's devoted dog refused, for the first time in its life, to feed from his hand when the day's work was done and retreated to its kennel early in the evening. The farmer collapsed and died shortly afterwards, though he had been in perfect health. At a public house a normally calm dog suddenly became frantic, tugging at its owner's sleeve and half-dragging him outside. Shortly afterwards the building collapsed, and the owner was convinced that the dog had saved his life. Nine died in the debris, and twenty others were injured. These and other tales have been claimed to show that animals can predict the future. It seems far more likely that they have senses which can alert them to problems before a human mind can recognise that something is wrong. Prescience must not be confused with prediction.

Nevertheless, there are countless examples which are hard to explain scientifically. A quarryman was always accompanied to work by his dog until one day, when the animal refused to go with its master to the quarry. Shortly afterwards there was a major explosion. In another incident, a dog tried to prevent its mistress from driving in a borrowed car, even jumping for the keys as though trying to wrench them from her grasp. She set off, and shortly afterwards was killed when the car skidded into a wall. Another pet dog of a holiday-making family became increasingly agitated until the driver was obliged to pull over and let the dog out of the car. Only then did they discover that the front wheel was in a dangerous condition and could have fallen off. The owner of a dog, itself crippled for years by disease, was suddenly seized with a conviction that the dog would be able to walk again. She telephoned the veterinary hospital, where her pet was under treatment, to be told the animal had indeed suddenly improved and was trying to take its first steps. In London, a pet Alsatian suddenly became agitated, and then sullen and depressed, at the very moment when its owner suffered a fatal heart attack far away in the tropics. There are many other vivid accounts of pets which have marked the death of their distant owners by sudden changes of behaviour.

These are isolated stories, and are not amenable to scientific analysis. It has to be said that belief in a telepathic sense is nothing new. Most of the recorded examples concern human, rather than animal, responses. Many university departments have looked into the matter, but so far no overwhelmingly convincing case has come to my attention. The question is simple, though it is not one we can address in this book: is telepathy real? If it is, we might expect animals (as well as humans) to experience the phenomenon. The examples in the literature do not confirm the genuine nature, or otherwise, of a supposed telepathic sense, but they do remind us of the interconnections between humans and the other creatures in this world.

Animals in communities

I have presented many examples of the universality of sense and communion in the living world. Some creatures reveal these imperatives in unexpected ways. Mole rats, comprising several different types of rodent, show a form of community which is astonishingly reminiscent of the organisation in insect communities, like bees and wasps. The mole rats are highly specialised for life underground. They are adapted for digging, with big teeth or strong claws and no external ears. Most have very short fur. The teeth in some species are the main facility for digging. Some have incisors so large they cannot be enclosed by the mouth, and remain projecting even when the animal is asleep. Those species that dig with their teeth have a flap of skin which closes the throat during digging to prevent soil from being forced into the animal's stomach. There are three families of mole rats: the Spalacidae, Rhizomyidae and Bathyergidae. The Spalacidae are all blind and tail-less, living around the eastern Mediterranean and inland as far as the Ukraine. Members of the Rhizomyidae family are found in East Africa and South-East Asia. The third family lives in Africa where some of the species are collectively known as blesmols. Of this group, the most remarkably adapted is the naked mole rat, *Heterocephalus glaber*. It lives in East Africa and is a naked rodent 15 cm (6 in) long. There are about two hundred individuals in a single subterranean colony, and only one female, the queen, is fertile. She produces the young and nurses them, while attendants care for her. If there is an attack on the burrows, the attendants seize the young and hide them, returning them to the brood-chamber when things have calmed down.

The bulk of the remainder are workers which fit into a strict caste system. There are poorly developed labouring mole rats, and above them are higher-caste animals which are slightly larger. It seems that the young which grow fastest are accorded a higher place in the hierarchy and develop further as adults. In the highest caste are those which care for the queen. The young females do not become fertile as long as the queen is healthy. She suppresses their sexuality. Some commentators believe she does this through potent secretions in her urine, while the view has also been advanced that it is her aggressive and dominant behaviour which causes the sexual repression.

This social structure depends on complex systems of signalling and communication. Each animal knows its place, and they all know what they are expected to do. The 'inner court' which care for the queen spend most of their lives in idleness, but if there is the threat of a raid on the colony they become highly active, ready to attack any invader, and rushing through the nest uttering shrill warning cries. It is remarkable how closely this strictly regulated community parallels the structure of the honey-bee colony (see Chapter 4). We see universal imperatives emerging in these very different groups of animals.

Other communities seem to reveal resonances of human society. One of the most intriguing communal creatures is the prairie dog, *Cynomys*. These mammals maintain complicated city structures which they organise and run with great efficiency.

If we look closely at these entertaining creatures, we find so many echoes of the way our own communities are organised. There are very sophisticated methods of communication to underpin this form of social structure. Of the several species of prairie dog, *C. ludovicianus* is best known for its extensive 'towns' built on the Great Plains of the USA. These endearing animals are members of the squirrel family, the Sciuridae. Each weighs about 1 kg (about 2 lb) and measures about 45 cms (18 in) in length. Their underground lifestyle has given them strong claws and small external ears, which do not get in the way when burrowing. One of the main dangers comes from eagles and buzzards attacking from the sky, so prairie dogs have their eyes unusually high on the head. When they lift their heads cautiously from a burrow to look around, it is the eyes that appear first, almost like a crocodile rising from a swamp. Living underground gives the prairie dog an ability to live in conditions with very different climates. For example, *C. ludovicianus* is found in North Dakota, where there can be heavy snow and severe frosts all winter, but other colonies live right down in central Texas, where the summers are blistering hot and severe frost is almost unknown. The factor common to all the communities is a level of annual rainfall around 50 cm (20 in), sufficient to sustain prairie grasslands and the creatures that live in the soil beneath them.

Like many humans, prairie dogs are homesteaders and cultivate their land. This they do, not by digging, but by selectively eliminating unwanted species. They cut back the tall grasses in the areas around their towns. This removes cover which predators would like, and it also frees the land for the growth of shorter annual species of plant which tall and shady grasses normally inhibit. There are many quick-flowering species around a prairie dog community, which provide the seeds on which they like to feed. If non-food plants begin to grow, they are often nipped off at the base and left to wither in the sun. The prairie dogs know which plants they like to encourage. Many of them are rich in water and since these animals rarely drink water, the plants are vital for survival. The prairie dogs are also landscape gardeners, and construct mounds around each tunnel entrance. These are about 60 cm (2 ft) tall and up to 2 m (6 ft) across. They are made with the soil from the excavation, and are maintained by the animals, which push soil up onto the mound when it has rained and the earth is soft and manageable. The mounds serve three purposes:

o They are a dump for excess soil from the burrows.

o The raised platform gives the prairie dog a vantage point from which to watch for predators, or for raids by prairies dogs from other towns.

o When the plains flood in heavy seasonal rains, the raised tunnel entrance prevents shallow flood-water from running into the subterranean chambers.

Eagles are always ready to attack from above, marauding badgers dig in from the sides, and coyotes can attack from ground level, so vigilance is an important component of the prairie dog's existence. Not all animals are sought out and repulsed, for the prairie dog can tell friend from foe. Some creatures use the burrows as a temporary refuge. Mice and rabbits have been known to take shelter in a prairie dog burrow, and they are usually tolerated.

As in human society, and elsewhere across the breadth of the animal world, the reactions of the prairie dogs are not simply predetermined. They learn from experience, and adjust their behaviour accordingly. If deer start to feed in the area of a town, they may well be the subject of intense interest by prairie dogs keeping watch. The deer, being larger and used to the ways of the world, take little notice. In time the inquisitiveness dies down, and once the prairie dogs are confident that the deer are doing no great harm they learn to tolerate their presence and will even feed very close to the deer.

Bison find the burrow entrances attractive as dust-bowls and like to wallow in the hole, churning it with their horns. The prairie dogs soon learn to live with their gigantic neighbours, and will feed alongside the bison. If damage is done to a burrow entrance the prairie dogs will come out in groups to carry out repairs. This is not an automatic or obsessive behaviour pattern, however. Sometimes the bison will find one or two sites particularly attractive, and will use them as dust-bowls time and time again. Once the prairie dogs see that the bison mean business, they abandon the entrance and waste no further time trying to restore it. They clearly have a reasonable understanding of the situation.

The interpersonal behaviour of prairie dogs is appealing to human eyes. They play together, rarely antagonising one another, and often touch or nuzzle face-to-face almost as though kissing. Within a single community this form of behaviour is the norm. The animals run into one another's burrows and everyone knows everyone else. These communities comprise 'wards' or 'chapters' within the greater town. Each is known as a coterie, and consists of one or two males, three to five females, and up to thirty youngsters. They live as an extended family and support one another. Each coterie occupies a different region of the town. If an animal wanders into the territory of a neighbouring coterie it is soon seen off. The daily patrols through the extensive underground burrows seem to be concerned with maintenance, and checking to see that all is well. The prairie dog does not have a ritualised habit of exploring set pathways. Observations confirm that, if one part of the town has not been patrolled one day, it will be a focus of attention early on the next.

The interaction between coterie members seems highly affectionate. Two feeding prairie dogs which come into sight of each other may run together to meet, or one may run across to sit with the other. When they meet they turn their heads towards each other, open their mouths and seem to kiss. Sometimes the kiss is prolonged, and one of the animals may roll onto its back, still kissing the other. Eventually they may both lie together for a while, and then move off to feed, staying close together as they go. There is much olfactory information exchanged during such bodily proximity. However, the long duration of this kissing and closeness must be because it is pleasurable for the animals. They are interrupting their feeding to indulge in this communion, and feeding is one of the most important drives all animals possess. There are interesting parallels between the prairie dogs' communication of their friendship and patterns in humans.

Sometimes two prairie dogs are not certain of one another's identity. If this happens, they do not approach each other immediately, but crouch low in the grass, watching each other intently. They approach slowly, bellies on the ground, flicking their tails in warning all the time. When they meet a sniff and a kiss provide the assurances required. Because the kiss begins with an open mouth (teeth exposed), ethologists have concluded that this is, in effect, a threat to attack by biting. To an unwanted intruder, the gesture is accepted at face value. If the pairs are members of the same coterie, the open mouth becomes a kiss and an exchange of friendship. That is a convincing explanation, but just because an open mouth is a sign of aggression it does not necessarily follow that this is the root of the gesture in prairie dog society. It may be a sign of aggression in humans about to attack each other, too, but we, as humans, have little difficulty in distinguishing aggression from a kiss.

The kissing of the prairie dogs is only the first of many stages of physical contact. After the kissing come grooming and petting. The animals nuzzle each other and one will stroke the other with its paws. The recipient lies in the grass and, if the attention slows down or stops, wriggles towards the other and nuzzles it until the grooming starts afresh. There is no mistaking their liking for this activity. There is no strict hierarchy for the behaviour: males and females groom each other and the youngsters; the young prairie dogs groom each other and the adults. The pups crawl under resting adults and nuzzle them repeatedly to receive grooming attention.

It is the call of the prairie dog that is its most noticeable means of communication. This is the origin of its common name. *Cynomys ludovicianus* is a member of the squirrel family, no more related to dogs than is a mole. Its cry, however, is vaguely like the 'yip, yip' of an excited puppy. There are three main types of call uttered by the prairie dog. The normal bark is a high-pitched 'yip' which can take many forms and is differently sounded according to the occasion. It may attract attention, for example. This is a basic sound with shades of meaning. Louder and repeated several times it is the penetrating territorial call. A third cry is the higher and more rapidly repeated bark used to warn of impending danger, or the approach of a possible intruder. The greater the hazard, the more intense and rapid the warning; the more urgent the sound, the faster the other prairie dogs respond.

The territorial cry is uttered as two vocalisations, syllables if you like, made loudly, clearly and in a sequence of two or three closely separated cries. It is delivered with gusto, the animal standing up on its hind legs and almost leaping upwards with the force of delivery. Youngsters often watch the performance and try to imitate it. With their forepaws outstretched they deliver the sound as loudly as they can, gathering themselves on their hind legs and throwing themselves into the business so heartily that they sometimes fall over backwards with the effort.

What do the prairie dogs achieve through this ritualised calling? Versions of the call are used by an adult male to remind others of the extent of his territory or to catch the attention of distant coterie members. The cry attracts the attention of a stranger, warning it of the consequences of territorial encroachment. In a slightly different form it is uttered as a bark of victory at the end of a successful encounter with an intruder, and the rest of the coterie joins in a chorus of triumph over a vanquished foe. Territorial calls can be heard across the entire prairie dog town. Each coterie leader has his own announcement, and coteries do not invade one another's territory.

There is much show of aggression when a prairie dog rushes towards an intruder. The two charge at each other and stop when face to face. One of the prairie dogs turns his back on the other and exposes his anus. He raises his tail, and the other prairie dog sniffs at his exposed anal glands, which secrete a characteristic scent. After a moment or two the roles are reversed, and the other prairie dog has a chance to scent at his opponent. This behaviour is repeated several times, until one of the two takes a gentle nip at the other's rear end. There is an indignant response, which ends in more repeated smelling encounters. Eventually, after much mutual hostility, the two back off and resume grazing. The encounter is a highly stylised ritual which does no harm to either party, and allows territorial boundaries to be reaffirmed.

Occasionally, a prairie dog becomes so emboldened that it creeps over to the entrances to a neighbouring coterie and makes a loud call into the burrow. Within seconds a defender appears and the intruder runs back to his own territory. It is reminiscent of a child knocking at someone's door and scurrying off. Invasions do occur, and they are a serious matter for the prairie dog community. They often result when a trespasser is not challenged on calling down a rival's burrow. Once in a while, one of these miscreants will invade neighbouring territory and stake a claim to it. It begins innocently enough, with what seem to be chance incursions over the territory occupied by neighbours. The invading prairie dog may habitually graze just beyond his rightful territory, and if the other coterie does not respond he will become increasingly bolder and stray further into the foreign ground.

If the neighbouring prairie dogs respond aggressively and issue a warning, there is every chance that the invader will turn on his tail and go home, even if first indulging in a little facing-out exercise with the neighbours. If they do not warn him off he becomes ever bolder, and one day refuses to leave after a visit. At this stage there is a fierce confrontation between the two males and the fight continues until one is clearly vanquished. From then on, the territory is defined afresh. If the invader is sent packing, he stays at home in future, but if he succeeds in occupying a new sector of the town the vanquished coterie retreats.

Coterie society confers many benefits on these animals. Their combined activities allow the growth of seed-bearing plants in areas normally overgrown by grasses. The efforts of watchfulness against predators are shared, and vocalisation makes sure that everyone in the town knows what's going on. The breeding season comes between March and May, and then the coterie system is temporarily abandoned. Tunnels and passages normally used as rights of way by members of the community become no-go areas, as females give birth to their young and suckle them. If a mother prairie dog is grazing in the open, she will contentedly associate with other coterie members, but within the underground chambers she vigorously defends the privacy of her own territory.

Driven out of much of their home, members of the coterie take this opportunity to extend their network of passages. They forage in more distant areas, and start to dig burrows outside the edge of the town. They don't sleep there, usually returning home at night, but this is how the colony begins to expand. Once the females have become fertile they typically raise a litter of about five young for four or five consecutive years. However, a propitious season can greatly increase the number of pups raised to adulthood. To prevent overcrowding, emigration takes place. It is at this point that the new burrows on the town's edge become important.

The new burrows become new suburbs reaching into virgin territory. If the coterie has become too large, some of the older members take charge of the new development and eventually start to spend nights in the new tunnels. The young pups make incessant calls upon their elders, especially for physical contact and endless grooming. The adults can become tired of the demands, and sometimes simply walk away and refuse to respond. These are the adults that begin to explore other, quieter, territories. There is no sudden desertion of the old homestead, for the transfer takes place gradually. As time goes by, a fresh colony starts to emerge in the new tunnels, and soon becomes established. On other occasions a well-established area is suddenly taken over by an expanding coterie group. This ordinarily happens in an area of tunnels where there has been limited activity, perhaps a part of the town in which the neighbouring coterie has been showing little interest. The invading group will make their presence known and, more often than not, will frighten the occupants away by weight of numbers. If the burrows are under-occupied, or the occupants do not assert their presence with an adequate display, they may find that their territory is taken over permanently by the dogs from next door.

As in human society, there are set stages in raising a pup. Once the youngsters leave the brood-chamber where they have been suckled by their mother, they encounter the father and other siblings in the freedom of the outside world. The adults are happy to groom the new pups, and the females tolerate them spending time sleeping in other litters. For the pups it is a delightful time. They kiss all the members of the community, each of whom responds. Mother still suckles them, and so do any willing foster-mothers. The males will gently discourage them if they attempt to suckle from one of them, and will lay them down to groom and caress them instead. They may even ruffle the youngsters' fur playfully with their teeth.

The young may wander off exploring, even entering a forbidden part of someone else's domain. Nobody seems to mind. As the weeks go by, attitudes change and territorial enforcement becomes apparent. The young trespassers are warned off at first, and may be vigorously rebuffed if they persist. It is much the same in their home territory. The pups find that the adults will spend much time grooming and playing with them. If an adult is unwilling, the youngsters may climb on it or squeeze under it, pulling at it to encourage it to join in the fun. With increasing maturity comes decreasing tolerance, and the fast-maturing young soon find that they are ignored or even pushed to one side by the increasingly disaffected adults. The young prairie dog starts to vocalise and to call to the others, who give a quick response. Sometimes they may wander into alien territory and try a call from there, but they soon learn that strangers are unwanted and future exploration is therefore confined to the area of their own coterie. The young prairie dog learns steadily from experience, and from the examples of others. In time, it changes from a carefree and tactile youngster into an adult with responsibilities and territory to protect.

Closer to humans

Of all other creatures, the monkeys and apes are the most like us. They share our senses, and 98 per cent of our DNA. As their behaviour illustrates, they also share many of our feelings. Baboons have much in common with us. Most of a baboon's life is spent close to other baboons. The community, known as a troop, patrols its familiar territory and the individuals support each other in many ways. The group is controlled by adults who are highly interactive and emotional, with individual personalities and recognisable habits.

Baboons live in troops of about sixty individuals. They are not united by habit, by pre-programmed subconscious urges, or even by sexual activity. Baboons form their communities largely on the basis of personality and emotional ties which bond them together. The prospect of a 'lone baboon' is unthinkable, for they seek and need continual interaction with their fellows. It is communion they are looking for, not sexual gratification.

When humans come along, baboons react with interest. They will climb onto vehicles and allow people to accompany them. Experienced observers soon get to recognise them as individuals, and claim that they have personalities as distinctive as those in human society. The birth of young is a matter of great interest to the troop. Many of the baboons come to see the new arrival, and they spend time grooming the mother, and trying to groom the newborn infant. The baby seems to be the centre of everyone's interest. Even young baboons come to see what the infant looks like. The adults have long-standing friendships quite apart from their mates. Female baboons form platonic pairs at an early age and these relationships frequently persist throughout adult life.

The newborn baboon is with its mother twenty-four hours a day. It has one important reflex from birth: the ability to cling tightly to the hair on its mother's chest. This is where it needs to hang for survival. The mother will lift the baby with her hand, but the clinging reflex is vital if the baby is to stay in place as the mother moves off or escapes from a threatening situation.

Social behaviour, however, requires learning. If the baby is reared in ideal circumstances and with great care – but in isolation – it does not learn the retinue of behavioural norms that young baboons ordinarily acquire.

Baby monkeys show an instinctive attachment to their mothers. Rearing them by hand, away from their mothers, may give them a safer upbringing with a lower mortality than in naturally reared babies, but it denies them the contact with their biological mother which their minds clearly need. They gain much from this contact, and learn much from their mothers. Harry Harlow carried out a series of experiments with hand-reared monkeys. He presented them with an artificial mother in the shape of a furry toy with a clearly recognisable face. Alongside he erected another dummy, this time made of a wire frame on which it was easy to climb, to which was attached a latex teat supplying formula milk.

Tests with wire models, some covered with furry cloth, showed that the young monkeys responded in ways that were physiologically identical. The baby monkeys fed just as easily from either, and the rate at which they put on weight and the time they spent over feeding were comparable in each case. Normally, it is only the physiological results in animals which we measure in science. What of the psychological? Here there was an instant difference between the babies' responses. They spent far more time clinging to the soft mother-substitute than on the wire models. Although the cages were warmed with an electrically heated floor, as soon as the baby monkeys were old enough to move on their own they would climb up to be with the soft toy.

There was no hesitation in feeding from the wire model. The baby monkeys were happy to suckle from the teat, and were perfectly well nourished by the milk with which they were supplied. However, once feeding was over they would always prefer something soft, and headed for the furry dummy mother for comfort. They would explore the toy face with their hands and nuzzle against it with their lips and cheeks. In itself, this is a highly revealing demonstration. It has often been taught that babies develop their feelings for their mother because she is associated with feeding. This result should lead us to doubt such a convenient, behavioural explanation. Baby monkeys need their 'mothers' to feel like mothers. Feeding has nothing to do with it. Though the furry model was impersonal, it provided some advantages for the rearing of baby monkeys. It was always available, twenty-four hours a day, even as the baby grew older. It never lost its patience or scolded the baby. No time was wasted in its own grooming, feeding or toilet requirements. It didn't drop or crush the baby, either, so the survival rate of the fostered monkeys was 100 per cent.

Mothers support their young in times of tribulation, and some tests were conducted to see how the babies responded in a situation which frightened them. The experimenters used a mechanical toy in the form of a walking clockwork bear beating a drum. As soon as it came too close, the baby monkey fled for the comfort of the mother's arms. Sometimes a young monkey would react so suddenly that it would leap onto the wire frame model and cling there momentarily. Once it realised that there was an alternative nearby it would jump from the wires and flee to the furry arms of the toy, as though seeking a mother's protection. After a few moments of security it would turn around and face the mechanical toy, now with greater confidence than before.

Young children are threatened by unfamiliarity. Put a small child in a strange situation and it will be content as long as it can see its mother, but the loss of visual contact will soon cause it to become upset. A similar effect is observed in the little monkeys. In one of the tests they were put in a room with several interesting objects: a doorknob, wooden bricks, a toy tree, a piece of crumpled paper. After a brief inspection the monkey would retreat to a corner and huddle uncertainly against the floor. Peering from between its arms it would look at the random objects.

All this changed if one of the objects was the furry toy representing mother. The monkey, faced with the strange objects, would flee to its protection and cling there for a while. It would clamber over the body and explore the face. Thus reassured, it would start to examine the objects, though with frequent glances back to the substitute mother. It might play with a toy, and then take it over to the mother figure, or inspect a wooden brick, and then go back to sit with the mother figure for a while. The little monkey was waiting for an adverse response from the mother. As long as she seemed to sit there, not minding, then the young monkey didn't mind either. Babies raised with nothing more than the wire-frame mother figure derived no comfort from its presence. If they were placed in a room with unfamiliar objects they would retreat to a corner and hide, their arms draped over their heads. The monkeys whose furry mother substitute was merely absent were less distressed in this situation than babies raised with no furry mother substitute at all.

Of course, the cloth-covered artificial mother did not move. Tests were performed which gave babies the choice of a stationary model, or one which was fitted with a rocking motor. All showed a preference for the rocking mother substitute over the immobile one. All the monkeys had access to a lever which they could operate to give them a view of a monkey outside, or even of a furry mother substitute. They were more stimulated by a sight like this than they were by the sight of a bowl of their favourite food. Was warmth a factor? No, since the little monkeys would quickly abandon the (heated) floor of the cage to cling to the (unheated) mother substitute. Is food the reason why monkeys cling to their mother? Clearly not, since the monkeys showed no preference for being fed from a wire frame or a furry model. How important is visual stimulation? It matters at a set age. At about three months, monkeys start to explore the face of the substitute mother, peering at the eyes and trying to manipulate the head.

Finally, how important is early contact with a mother figure, even a surrogate one? Some of the monkeys were raised in conditions of warmth and abundant food, and at the age of eight months they were given wire models and furry surrogates. They were initially somewhat afraid of them both, but soon accepted them as objects worth investigating. After a week to become accustomed to their new playthings, they would spend less than an hour a day playing with the wire frame, but up to ten hours with the fur-covered mother substitute.

More interesting still, these orphans spent less than half as much time with the cloth mother substitute as did the monkeys raised with it. This is a highly significant finding. Some of the little monkeys were left with the surrogate mother up to the age of about five and a half and were then separated. When they were reunited with the furry model, up to eighteen months later, they showed immediate attraction to it. Some even seemed more attentive after the period of separation. It was clear that the early experience had provided sufficient input to establish a bond. For the monkeys which had never been with the furry model, the pattern was very different. When introduced to the model for the first time at eight months, they responded to it positively. However, not only was the amount of time they spent with it much reduced, but they soon lost interest altogether. The monkeys denied the chance to become acquainted in their early weeks could never develop the closeness at a later age.

Another experiment in an American research centre offered

rewards to a monkey in captivity in return for its performance of specific tasks. One of the rewards it valued most highly was the sight of another monkey. That alone indicates the strength of bonding which can occur in these creatures. They have strong feelings for each other, and go to great lengths to nurture and preserve their relationships.

This might well apply to neglected human youngsters who receive little parental input at appropriate ages. Our era tends to reject children. Parents are 'stuck' at home with them, they are seen as 'little monsters' or 'nerds' who are devoted to little other than their computers. Many grow up feeling unwanted and rejected. If this is the psychological input they receive at a key stage of their development, it is easy to see why their attitude to society as young adults is non-participatory at best, and destructive and violent at worst. We are what we learn. With so many warnings in the media about fatty foods, the dangers of over-eating and continual reminders of the percentage of the population construed as overweight, we should not be surprised when young girls retreat into anorexia. They are doing what the adult world has taught them. Experimental work with primates has confirmed that parental input and sequential learning are vital components of the maturation process. This is a lesson that Western society has yet to apply to the young we raise at home.

Baby monkeys in the wild are given attention not only by their mothers, but also by others in the group. They have instincts which compel them to cling to her furry body, but the socialising they need to acquire is learned. So is the way they have to travel. From the early position of clinging to the mother's chest, the young baboon quickly discovers how to ride on its mother. She lifts the young monkey up and seats it on her back, where it quickly learns how to cling in this new position. Quite soon it learns how to eat solid food, and also how to play. It is introduced to other youngsters and they interact as a group. There is much teasing, tail-pulling and play-fighting. An adult male is usually close by to keep an eye on what goes on, and if a youngster is hurt in the games the intervention of an adult soon restores harmony. Grooming becomes an important part of the baby baboon's life. They obviously enjoy it. As the baby matures, much of its grooming is done by its mother. The adult females do most of the grooming in a troop of baboons: they groom their young, they groom one another and they also groom the adult males.

Several purposes are served by grooming. One is basic hygiene. The grooming allows baboons to pick out specks of dirt, and they also find parasitic mites or ticks. The fingers, and the teeth, are frequently used. Another purpose is cosmetic. The action works like combing the hair, and the matted or tangled fur is made smooth and clean-looking. A third value is the social contact and the friendship that results. Baboons, like many other mammals, groom in pairs. The animal receiving the attention leans backwards and forwards, head raised and eyes closed, very obviously enjoying the experience. After a while the two change roles. The former recipient of the treatment now gets to work, diligently searching through the fur of their partner, ceaselessly stroking their body with curved fingers. The main function of the grooming phases, which occupy much of the baboons' daily routine, is the sheer pleasure it gives. These animals enjoy the benefits of manual contact, much as we like massage. There is nothing mechanical or automatic about all this; they do it because it's sheer pleasure.

Baboons have a strong sense of community and avoid encroaching into areas occupied by other troops. This is a social constraint, not an overwhelming instinct. If water is short, for instance, different troops may intermingle at a watering hole in a manner which would otherwise be unthinkable. Although baboons are largely herbivorous, they will sometimes kill and eat other mammals. However, they happily co-exist with very different species. Baboon and impala are frequently found feeding together. The impala, tall and graceful, have the keen eyesight of herbivores and respond quickly to an approaching foe. When a troop of baboons is with them, a herd of impala becomes bolder. Prowling cheetahs do not frighten them. A male baboon can send the cheetahs off with a fierce growl and a threatening posture, while the impala carry on grazing, unconcerned. If a baboon sends out a warning cry, then both the baboon troop and the herd of impala will respond in an instant, and flee.

These communities of prairie dogs and baboons are just two examples of mammalian social systems. They are not animals held together by lust, or forced to cooperate through some irresistible instinct. Animals consider options, learn from one another and from experience, find new solutions to sudden problems. They feel pleasure, fright and anxiety. We celebrate our emotions, most notably in cultures where psychotherapists are fashionable. Animals also live with emotions. It is time we understood them, rather than denying that they can exist.

Orangs as delinquents

The orang-utan is a beautiful and responsive creature, and there has long been a flourishing black-market trade in young orangs as pets. They are a protected species, and to prevent kidnappers from taking them, large numbers of the young have been removed from the wild and reared in captivity by the Indonesian authorities. Raising these creatures is a costly business, and in recent years attempts have been made to restore them to the wild. Choosing protected sites on isolated islands (where they used to abound), the government advisers have set up centres where the orangs can be acclimatised to life in the open air prior to release.

A new study by the Fordham University in New York State shows that the policy does not succeed. Raised without the full learning experience of the natural environment, the captive orangs do not know how to care for themselves. They lack social skills, and cannot undertake the tasks they need to survive. Rather than learning the ways of the wild, the captive creatures bond more to their human keepers. Once they have been released, they act aggressively towards orangs they do not know, and rely on feeding from the release centre.

The Fordham University research centred on 27 young adults which were released into the Tanjung Puting National Park in the early 1980s. The normal life-span of an orang is about 60 years, so there was clearly a life expectancy of decades in prospect. More than half of them soon died. At an estimate, there may be seven or eight still alive today, perhaps fewer. These are orangs which were healthy, had lived active lives and were reared in idealised surroundings. Every aspect of their artificial environment was the result of the scientific study of what they eat and what they need to remain physically fit.

What it could not do was give them the care and teaching of a parent. Young orangs take at least seven years of daily teaching to become fully functional individuals. Their mothers teach them how to forage, how to identify foodstuffs, and how to acquire the skills of life in the forest. They become socialised and learn life skills. These attributes can only come from a parent, and can only be learned in the wild. Without them the orangutan is a hopeless victim of its surroundings and cannot master its life. They return to feed from the shelter as often as they can, and if they travel too far away they are likely to die. The effect of the learned behaviour is so important that orangs from different regions acquire different cultures, and cannot form communities. They are kept apart when raised in captivity.

What is the future for these marvellous and entertaining creatures? The only chance is for them to be maintained in the wild, not in captivity. Communities from zoos have lost their ability to survive without support from people, and the most sensible answer will be to release them on islands where they can continue to have human support. Visitors could flock to see them as eco-tourism takes hold. They would be in their ancient environments, but reared by human intervention. One lesson is clear – these creatures must learn from their parents, and if they do not do so their community culture is lost forever. Wild creatures must be raised in the wild; the zoo can never successfully replace it. The need for the learning of social skills is as vital for wild animals as it is for our own young and, although zoos can conserve creatures, it is the living culture of life we need to protect.

Apes and humans

Apes, as social animals, show many of the traits we like to think of as human. For many years the popular view has been that humans descended from the apes. This stemmed from the response to the theory of evolution promulgated by Charles Darwin (1809–82) and Alfred Russel Wallace (1823–1913). Caricatures of humans with ape-like characteristics became popular in the mid-Victorian era, as evolution became a fashionable topic of conversation. More recently it began to seem likely that chimpanzees, gorillas and humans all evolved from a common ancestor, itself now extinct. The current consensus seems to be that chimpanzees are more closely related to humans than is the gorilla. There is genetic evidence that the pygmy chimpanzee or bonobo may be our closest relative of all; indeed one school of contemporary thought regards humankind as a third branch of the chimpanzee lineage. It would be from this single line that our cave-dwelling ancestors descended. Neanderthal man would have been replaced by the more highly developed Cro-Magnon man, and it is from this stock that modern man, *Homo sapiens*, would be descended.

It seems that the gorillas split off from the human line of descent about 10 million years ago, and chimpanzees went their own way about 5 million years later. Humans have existed for perhaps 2 million years. Negro and Caucasian stocks may have diverged as recently as 50,000 years ago, and the Oriental races may have diverged from the white Europeans as recently as 10,000 years in the past. That explains why all humans are the same species. It also makes us realise how close we are to our ancestors. If a human generation is taken as thirty-three years (which makes the mathematics easy, with three generations per century) then there are only 6 million generations since the earliest *Homo sapiens* of all. If people stood in a single line, shoulder to shoulder, the distance all those generations would spread is 4 million metres (over 13 million ft). That is a mere 4000 kilometres (about 2500 miles) – approximately the distance across the USA (4517 km, or 2807 miles) or Australia (4025 km, or 2501 miles). That is how close we are to our pre-human ancestors. Add enough relatives to stretch twice as far again, and we are back to the point where we divided from the line that developed into chimpanzees.

There are two species, the common chimpanzee, *Pan troglodytes*, and the pygmy chimpanzee or bonobo, *Pan paniscus*. Together they make up the primate family Pongidae. The common chimpanzee extends across a large range of Africa, from Sierra Leone and Guinea on the Atlantic Ocean to Lake Tanganyika and Lake Victoria in the east. The bonobo is restricted to the eastern Congo River Basin. The common chimpanzee has much in common with humans. Males grow up to 1.7 m (5 ft 6 in) tall. They weigh up to 70 kg (150 lb, or 10 st 10 lb). Like us, chimpanzees are diurnal and omnivorous. They eat about two hundred different kinds of vegetation and fruit, supplementing the diet with ants and termites, honey, birds' eggs, and small animals – including other monkeys from time to time. They are far from being docile herbivores, for they can outwit and work together to trap and kill monkeys as a supply of meat. They are ingenious builders, too, for the adults build a sleeping nest in a tree each night.

The female chimpanzee has a menstrual cycle lasting 35 days. She is sexually receptive for six days in each cycle, and she can mate and breed in any season. The gestation period is over seven months; as a rule a single offspring is the result, though twins are sometimes born. The young are weaned by the age of four, but they usually travel around with their mothers until the age of ten. A crucial lesson of this book is the way animals learn from their parents, and the chimpanzee provides a classic example. The young are painstakingly taught by their elders, and grow towards adulthood learning how to behave, how to act within the community, how to forage and feed, how to recognise food and herbs and medicinal plants that can help them treat illness. The young maintain a relationship with the mother throughout the rest of their lives. In the wild they live to be about sixty. In so many ways, chimpanzees are our distant cousins.

Chimpanzees form loosely organised bands of up to 80 individuals on fairly large home ranges, where the animals remain for years. Within a band, smaller groups may form, break up and re-form; sometimes a female migrates to another band. Males never migrate. Except between mother and young, there is little permanency in individual relationships. The female may mate with different partners. Members of a band cooperate in hunting and sharing of food. On finding a food source, they hoot, scream and slap logs to attract others. There is a constant interplay between adults, and all members of the group groom one another.

The eloquent ape

Communication is clear in the world of the chimpanzee. They make facial expressions which others interpret, and have a range of postural changes and movements, each of which conveys subtle meanings. A young chimpanzee can make more than 30 different sounds. Their emotions are often unmistakable, and they can show intelligence and evidence of thought. Chimpanzees share their food, for example. This altruism seems to connote thought. So does the observation of social behaviour. A notable observation was made of a chimpanzee named Dandy, who was courting a female in his group. A sexually excited male chimpanzee shows his erection to a female to indicate a desire to indulge in sex, and Dandy did just this to his chosen partner. At that moment, a more dominant male came on the scene. Dandy's immediate reaction was to drop his hands to conceal his penis. Does this mean that chimpanzees are shy? Their mating rituals do not suggest that they are. It seems much more likely that the chimpanzee has a concept of tactical response to a sudden situation. Surely this is a sign of thought? The fact that apes can be taught to communicate demonstrates some ability to think Some of them have such a good grasp of basic sign language that they can exchange ideas with their keepers.

The first ape to be taught American sign language was a chimpanzee named Washoe which was raised by Alan and Beatrice Gardner of the University of Nevada in the late 1960s. She arrived in North America when only a few weeks old, and has spent all her adult life in the company of humans. To help her start, the trainers would take her hand in theirs and gently form it into the signs they were using at the time. As a result she became adept at using sign language to communicate and understood her human keepers well. In her first four years

Washoe learned to use 130 signs, each representing a word or short phrase. She responded appropriately to objects or images. On one occasion, when she saw a swan for the first time in her life, she is reported to have signed 'water bird' to the Gardners. Washoe was also seen to sign to herself in quiet moments while relaxing in her enclosure.

At Kyoto University, Japan, a 21-year-old chimpanzee named Ai has also been taught to communicate using signals and signs. In 1998 she was reported to be pregnant (the birth being due in August 1998). Special attention is going to be paid to what she teaches her young. Since she now uses sign language as a convenient means to communicate with people, researchers are excited by the possibility that she may teach her young this human way of communicating.

The University of Pennsylvania has an ape which was taught to reply with coloured tokens made of plastic. She could answer questions like 'What colour is . . .?' responding correctly whatever the shape of the token, even if the colour of the token were quite different from the colour of the object it represented. At Emory University, two chimpanzees were taught to use a keyboard to communicate. They were soon able to answer simple questions put to them in the same coded language. Two of the males not only learned to communicate with each other using the system, but could actually ask each other for objects using the keyboard.

Could this be an elaborated version of the 'Clever Hans' phenomenon (see p. 44)? It is certainly true that many of the words and phrases communicated by the chimpanzees in American sign language began with exact copies of the signs made by their human companions. You would expect this, for they had been taught by copying in just such a way. Sceptics could argue that the chimpanzees were picking up subliminal cues of approval from their trainers as they responded using coloured tokens or pressed keys on a keyboard. Although that explanation is perfectly feasible, the complexity of the situation rules it out. There were so many signs and combinations that finding the right one by chance would be one in a thousand. In addition, safeguards were built in by using independent observers.

In Washoe's case, a trainer would sign to the ape while some-one else kept watch out of sight, behind a screen, and recorded the ape's responses. This reduced the chances of subconscious collusion, and eliminated any subjectivity in interpreting the results. One aspect of the sceptics' response is correct: the apes cannot say very much in sign language. They can string a few words together, but do not develop any sort of syntax or gram-mar. Young children use words in sequence, often adding a third or fourth word to amplify the meaning of one or two, whereas apes simply recite words in a scattered cadence that accords with their feelings at the time. A gorilla named Koko was taught sign language, and his longest 'sentence' was inter-preted as, 'Please milk please me like drink apple bottle.' A young male chimpanzee named Nim similarly signed, 'Give orange me give eat orange give me eat orange give me you.'

Although this is much less refined than the form of expression that a young child possesses, it does demonstrate the existence of thought in the minds of apes. One argument about language is that a signalling system deserves the name 'language' only if it can be shown that the individual receiving the message under-stands the mental attitude of the individual sending the message. This is not as restricting as it seems: a blackbird's warning cry is taken by others of that species (and by other species too) to indicate a state of alarm on the part of the bird making the sound. Thus, other birds do know what the sig-nalling blackbird means. So, come to that, do humans. The use of human sign language by apes also substantiates the view that words are irrelevant to thinking. Once taught sign language, these apes were able to use them to articulate their inner thoughts. Until humans instructed them on the conventions, they had no 'words' at all. Once they had been taught to use sign language, they used it as easily as any natural form of communication. This does not indicate merely that chimpanzees can be taught to communicate – they have always done that. These exciting experiments show that chimpanzees can actually *translate*. They have learned the language of another species, and can use that to express their wishes. Humans may have given them new words, but the animals have always had their language.

There are other lessons we must learn. A similar chimpanzee named Lucy was, at the age of eleven, taken from her human family and relocated in Africa, where she was released back into the wild. She was studied, from a distance, over the next three years. Unable to relate to the other chimpanzees, and lacking the knowledge she needed to survive, her life became a miserable existence and she plainly suffered as a result of this well-meaning gesture. The message is now clear: these creatures – and many others – need the attention and training they can only derive from their parents. Left to their own devices, or reared by compassionate humans, they lose their place in their own world and cannot recover.

Animals in captivity

Apes are not the only animals to have learned to translate a human language. Dogs and cats can hazard a guess at one's mood, and can respond to body language and voice-tone almost as though they understand the words. I know that it is popular to train dogs to obey specific, unambiguous commands (like 'heel!' and 'sit!'). However, animals can be acclimatised to understand commands expressed in unfamiliar words, if the tone of voice is unmistakable and the context clear. You can say to a dog, sitting on a forbidden chair, 'What do you think you are doing there?' and the dog will depart – even if that is the first time the words have been used by the owner.

Mammals are particularly astute at learning languages invented by us. Sheep-dogs are trained to recognise shouts and whistles, and can round up animals with astonishing skill. In the USA, Ronald A. Schusterman has taught a sea-lion named Rocky over 190 different gestures. This is one-way communication, but bottle-nosed dolphins can learn auditory or visual signals to carry out sequences of instructions. Naval trainers have shown that a dolphin can recognise symbols like 'bottom hoop, surface' and carry out the actions as required. These are sensitive creatures, highly attuned to their surroundings (and to us, if we are part of their world).

The realisation that animals are highly sensitive to their surroundings should make us look hard at how we care for captive creatures. The use of animals in scientific research is a key issue on which strong views have always been expressed. On a television programme I once discussed the subject, emphasising the extensive areas of medical progress in which animal research had been crucially important. Within a week there had been challenges to my view. One of them came from a correspondent with venom in her pen. She was, she said, opposed to all forms of cruelty. The world was run by cruel people, and she was convinced that the cruellest of all were the scientists who worked with animals. Cruelty, she said, had to end. She also had an answer to it all: the scientists should be strapped down and tortured.

The European Union has formulated a policy on our use of animals. We should be committed to 'the improvement of the welfare of animals from which we profit and for which we bear the responsibility of care'. The word 'improvement' is open to interpretation. Were we to be pedantic, we could argue that, instead of being excessively cruel to some animals, we should only be very cruel to them. That could be seen as an improvement, but does not eliminate the cruelty. We need a single criterion by which to judge our actions. It is simply this: we should avoid cruelty to animals. This is not only because the animals are sentient, but because humans have a moral awareness which should make cruelty inimical to civilised life. Animal 'rights' are not the issue. No mouse can wave its 'rights' at a hungry cat. Gazelles have few 'rights' over predatory cheetahs, and the right to survival by innocent villagers is rarely heeded by the killer elephants of Asia which set out to protect their ancient territories. I am more concerned with the duty of people. We are aware of the concept of cruelty, and it is incumbent upon us to avoid inflicting unkindness or suffering on our cousins. Biologists are devoted to encouraging a fuller appreciation of all life, developing the study of living systems, nurturing all wildlife and protecting the diversity of nature.

Biologists are also concerned with the promotion of health – not only human health, but the health of all plants and animals. Many diseases which afflict humans and other mammals are currently the subject of intensive study, including many genetic disorders. Objectors to the use of animals in research fall into several categories. Some, a fairly small percentage of an entire population, believe that we should not use animals for any purpose. Others, a similarly small percentage of the population, seem to feel you can do anything you like to an animal, because humans are ineffably superior. The views of most people seem to lie between these two extremes: they feel that the exploitation of animals by humans is acceptable, but only if it is constructive and humane. People in one of those schools of thought sometimes masquerade as adherents of another. The last time I spoke to a young student agitator about the subject was at a railway station buffet. He was adamant in his rejection of the use of animals. Indeed, he said he was largely vegetarian. I have no quarrel with the sincerity of such views, though he dented the impression of conviction somewhat by asking the buffet staff to pack his bacon baguette in a bag so that he could slip it into the pocket of his suede jacket.

Do animals feel pain?

Pain is a large subject which could fill a book on its own. Science still knows very little about human sensations of pain. There are different types, but we have no detailed vocabulary for describing them. Nor can we effectively measure pain, though the varieties of the sense can be categorised. Cutaneous pain, the pain felt by the surface of the body, is often localised and sharp. Deep pain, in the muscles or bones, tends to spread and can be hard to locate. Then there is visceral pain, which can be deeply disturbing. One of the commonest forms of pain is headache, and we all know how many different types there are. Some are throbbing, some searing, others are sharp; the pain may be diffuse, located in the temples, at the front, or related to movement. Although it is hard to describe pain, it is harder still to communicate the effect it has on us.

We have heard since we were children that animals (lobsters? fish? sheep?) cannot feel pain. There has never been any evidence for this bold assertion, which is usually a stance adopted to avoid the expense of treating animals humanely. Nobody has ever known that animals cannot feel pain. I have even observed microorganisms under the microscope which seemed to be showing signs of distress. Let's redefine that statement in more scientific terms: *there is no evidence that animals can feel pain.* Now, that is something we can examine objectively.

There are two components to the argument: can animals feel pain, and does it matter? The question of pain as an experience can be viewed from the standpoint of scientific knowledge. For example, does an animal respond to an unpleasant stimulus by showing an avoiding reaction? Many animals do this. Insects dying of a pesticide spray writhe and distort their bodies as though they are suffering. Beetles show a strong aversion to the presence of a hot needle brought near their antennae. Rats injected in experiments are said to flinch, and to learn that the experience is unpleasant. They learn to bite at the hand of the technician who comes to collect them from the cage. On this basis, animals can feel pain and can create a strategy designed to reduce its likelihood.

We can also ask whether there are sensory organs or cells at appropriate parts of the animal body. We can demonstrate the presence of pain sensors in a range of animals, and conclude that pain cannot be a rare experience in nature. Can we demonstrate nervous pathways that might transmit the pain stimulus? Yes; they are widespread in the animal world. Again, are there receptors for pain-resisting opium-like compounds in the animal? If there are, we could deduce that the animal was accustomed to dealing with the sense of pain. Finally, can pain-killers of known efficacy reduce an animal's response to a stimulus construed as being painful? This is certainly true; animals suffering pain which are treated with known analgesics show a reduced response to pain. It is widely accepted that mammals are highly sensitive creatures, but it is also said that cephalopod molluscs are sensitive too. They are said to recognise individual humans, and to avoid those that may have caused them pain in the past. There can be little doubt that animals feel pain, and there is a weight of scientific evidence to suggest that they do.

Does it matter if animals feel pain? In many cases the answer may be 'no'. Our anthropomorphism may well disturb our objectivity here. When we consider the sensation of pain, we do so in the sense that we experience it. Pain is a deeply disturbing

human affliction, but that is because we are aware of its implications. I have already referred to the way in which a deep cut
caused by stepping on a broken bottle on the sea-bed is likely to
be ignored while bathing, only to become painful in the
extreme once the wound has been noticed. For animals, on the
other hand, there is no consequential understanding of the significance of pain: it is often irrelevant or, at least, not traumatic.
Although insects may strenuously avoid a heat source near their
antennae, there are many that continue feeding on leaves while
being devoured from the rear. A mating male frog will continue
to hang on to its mate in a typical breeding posture, even if the
hind parts of its body are severely damaged or even detached.
Recent research at Cornell University on the male Australian
redback spider shows how its sexual performance is related to its
fate. The male spiders, after mating, are liable to be eaten by the
female. Up to six males will mate at any one time with a female
as though competing for status. The average length of time for
each copulatory act is 11 minutes, but the male which seems to
satisfy the female most of all maintains copulation for almost half
an hour. This is the contender she selects, and she shows her
appreciation by eating him, which prevents his genes from fertilising any other female. During the act the male spider shows
little inclination to escape.

Here we are considering cold-blooded creatures, but mysteries remain when we consider mammals and the response to
pain. You can find zebra grazing in African game parks after
being attacked by predatory big cats. With a great swathe of
hide torn from its flank, a wounded zebra will graze in a manner
that seems normal, apparently showing no response to pain.
Wild horses with a broken limb (an unendurable source of pain
for the typical human) will move about the prairie and feed
apparently without concern. The question of why a zebra with
a torn flank or a horse with a broken leg does not wince is fun-

damental. Is it because it realises that to show signs of pain would be counter-productive (being irrelevant to other horses, and a sign of weakness to predators) or because – through the blocking of pain impulses at the neuronal level or the production of pain-masking endorphins in the brain – it truly does not hurt? Experiences we regard as painful are widespread in the animal world, and we need to know more about their sensation of pain. It seems clear that they do not respond to pain in the same way.

Perhaps pain is not the problem. We may be asking the wrong question; it would be more revealing were we to ask instead whether animals suffer distress. The confinement of animals in zoo cages, or in cages as pets at home, can often cause them great anxiety. Animals can suffer from distress, insecurity or isolation (particularly separation from a mate) and can certainly react strongly to the threat of a situation they anticipate as traumatic. If we are to adopt a new attitude for the new millennium, it should be to reduce the burden of stress we cause to our fellow-creatures. This is certainly important at the human social interface and at the place of work, and I now believe it should also apply to the way we use animals. We have incalculable mental abilities, by virtue of being human. Now we need to show our humanity to the rest of nature.

Agricultural imperatives

Farm animals can show distress, and the pitiful lowing of milk-cows for their calves is well known to all dairy farmers. We consume farm animals in our diets, either directly (as meat) or indirectly (as milk, or meat products like the gelatine found in confectionery and biscuits). Meat products are widely used, and they are found in the least likely places: in wine gums and ice-cream, in marshmallows and liquorice. We use the products of animal farming in many areas of life; by-products like lactose and rennin, vellum, parchment and gelatine all play a vital role.

Farm animals remain an important part of rural management. For example, much of Britain's countryside is maintained through grazing. Management by other means would be impractical, and the resulting populations of farm animals form a resource which is valuable, and which also needs to be utilised. This maintains the pressure on the British people to continue with a diet rich in meat. There are no viable alternative proposals to utilise the animals, yet there are no apparent means of maintaining the countryside without them.

In recent years we have experienced a climate in which money mattered above all. This is always a mistake. Financial considerations are a crucial component of any stratagem, but must never be the main factor. In farming, the pressure to maximise profit (at any cost) has given us cruel and intensive systems of animal rearing. The managers of vast dairy farms clock on and off like office staff, and know little of the animals in their care. Pigs and veal calves are trapped in confining pens. The picture of the intensive farm is unpleasant and inhumane, and the over-use of antibiotics as growth promoters is leading to greater problems of human disease.

Much of this is unnecessary. Pigs on a free-range farm are a pleasure to see – roaming free, rooting in the ground or just resting in the sunshine. Farming need not be inhumane. There are moves under way to encourage the considerate rearing of farm animals. In London, the Royal Society for the Prevention of Cruelty to Animals has set up a 'freedom food' campaign. The RSPCA sets out five simple points which are used as operating principles in this approach:

1. *Freedom from Hunger and Thirst*. Animals must have ready access to fresh water and a diet to maintain full health and vigour.

2. *Freedom from Discomfort.* An appropriate environment must be provided, including shelter, and a comfortable resting area.

3. *Freedom from Pain, Injury and Disease.* Prevention should be the aim, with rapid diagnosis and treatment of problems.

4. *Freedom to Express Normal Behaviour.* Sufficient space should be provided, with proper facilities and the company of the animals' own kind.

5. *Freedom from Fear and Distress.* Conditions and treatment should be arranged so as to avoid mental suffering.

There are problems with campaigns such as this. Some enthusiasts might say that routinely administered antibiotics are part of a diet for full health and vigour, and argue that these precepts permit artificial farming methods. Opponents of farming might argue that any rearing of farm animals for slaughter must always involve some form of mental suffering. However, let us not lose sight of the fact that this is a formula which many producers and retailers have found acceptable. The 'freedom food' movement has gained adherents from many strata of society, and we need more activity of this sort to highlight the plight of animals in our care.

Agriculture has much to learn from the notion of sentience in farm animals. Intensive rearing of chickens and pigs is often too cruel to be condoned. A future society will not accept methods that insult an animal's need for what have long been called, with a certain prescience, 'creature comforts'. Cruelty in farming demeans the farmer, and often produces poor produce which damages the business of agriculture. A condition known as PSE pork (the initials stand for pale, soft, exudative) is related to stress before slaughter, and can make the meat unsaleable. It has long been thought that high-profit agriculture is a form of farming which maximises long-term gains. That is unlikely to be true. The financial problems posed to farmers by PSE pork are just one well-documented example. It was this attitude of short-term expediency that shaped the first official responses to bovine spongiform encephalopathy (BSE). Warnings by scientists were ignored, and in some cases suppressed. The result was a catastrophe for the business of farming. We are faced with similar problems with a lethal strain of *Escherichia coli*, type O157. Currently, as many as 10 per cent of cattle carry this dangerous bacterium. It does them no good, and it often kills the human consumer. The combined effects of *E. coli* O157 and BSE may well sound the death-knell of the beef business. If so, it is the hunger for a quick profit which has caused it. Well-nourished cattle raised in healthy circumstances are less liable to run into such problems. But the incidence of *Salmonella* in commercial chicken populations has already harmed the economics of that branch of agriculture. An attitude which places animal health and welfare high on the list of priorities will protect them from infections, and in the process will protect the consumer.

There is a current epidemic of a protozoan organism known as *Cryptosporidium* ('crypto' for short). It causes loose bowels in farm animals, and an epidemic of an organism which causes diarrhoea makes the animals excrete liquid slurry, which is easier to tackle than semi-solid cow-pats. *Cryptosporidium* has already caused some outbreaks of disease in humans. It spreads in liquid slurry when there are overflows or spillages, and gets into watercourses. Chlorination does not destroy it, and it can survive in tap-water. If we insisted that farm animals were healthy and well-cared-for, there would be controls on outbreaks of *Cryptosporidium* in farm stocks, and consequently fewer outbreaks of the disease in the human population.

Our current policies are in a state of bewildering confusion. For instance, people are not allowed to swim in artificial lakes used as storage reservoirs of drinking water because they might secrete pathogenic bacteria into the water. I recently visited one such lake. People are allowed to sail upon its limpid surface, but not to swim in any part of it lest they bring disease to the water. Meanwhile, a grass management scheme has been introduced, in which sheep are herded around the edge of the lake. Each day a single sheep can secrete enough *Cryptosporidium* to infect a moderate-sized city. One day, someone will sue one of the water companies. Once again, a poorly thought-out scheme threatens the future of a major enterprise.

The cramped battery cage has every appearance of being inhumane, with chickens crowded in with no more floor-space than a sheet of notepaper. That is not to say that we can merely abandon the cages and turn the chickens out, for they can suffer from suffocation if reared in deep-litter barns and even cannibalism if reared in the open air. Our future strategy needs to be based on sound research and thorough planning. In some areas, immediate changes could be made. The long-distance transportation of animals before slaughter should be ended, for they suffer when crowded into trucks. Slaughter near the farm where they were reared should always be the norm. In Britain a particular problem was posed by veal calves. These animals, routinely taken from their mothers in the dairy industry, were of little value in Britain but if they were raised according to practices in continental Europe they fetched a high price. The financial imperative drove the industry. Once the scale of the BSE epidemic emerged, the exports were curtailed and the entire business collapsed.

We also need to consider the future shape of agriculture. Many farm animals would benefit from the provision of some form of shelter in a corner of a field. It could offer a sense of territory which could aid productivity and perhaps even improve meat quality. High-efficiency farming can flourish best if it keeps animal welfare in mind. Schemes that are poorly thought out, and which do not prioritise the humane treatment of the animals, could destroy the business altogether. We are threatened by diseases spreading from the animals to ourselves: make the farm healthy, and human health follows like day follows night.

On the design front, I believe that hard and unyielding surfaces should be eliminated wherever possible. Animals do not like them, and most do not encounter them in nature. Veterinarians ought to move towards soft tables, rather than hard surfaces, for the examination of domestic animals. Soft litter should be used when animals are to be treated by a veterinary surgeon (I have seen zoo animals wounded when they fall after being anaesthetised, which a litter-bed would avoid). Animals do not require molly-coddling, but we should bear in mind the conditions under which they evolved before we design facilities for handling them in the modern world of technology. There should also be a new approach to animals as pets. Some of them, raised on over-indulgent diets, with poor exercise and a restricted habitat, exist under conditions of desperate (if unintended) cruelty. There is already a move towards more humane methods of training horses, and animals can grasp the sense of instructions from the human trainer without recourse to the obsessive use of a single word for every command. Domestic dogs, for instance, can become used to the whole range of human body language rather than being conditioned, like machines, to strict verbal commands.

Our self-centred view has encouraged us to imagine that human language is the only one worth knowing, and that mammals acquiring some response to body language are beginning to master language for the first time. I would like to persuade you that this is demonstrably untrue. Mammals have languages of their own. They communicate with each other on many levels, and transmit inner feelings – aggression, fear – to their fellows. The common nature of these language signs means that we can understand many of them. Nobody is likely to mis-interpret the fact that an angry rottweiler is adopting a threatening posture, or that a wounded terrier is distressed. Yet behind this lies a remarkable fact. Many mammals have learned to interpret the language of another species. A horse interacts with the wishes of the rider. Pet dogs can respond to emotional displays of the owner. Most remarkable of all, some primates have enough mental ability to learn human sign language. Those who imagine that this is the beginning of mammalian language are far from the truth. They have had languages of their own since long before humans evolved. Now, they are actually beginning to embrace one of ours as well.

Animal doctors

Some animals have an ability to seek treatments that may help to cure their ailments. The term *zoopharmacognosy* was coined at Harvard to cover the study of animals that can seek out their own medication. Salt-licks are perhaps the best-known example. African animals have long been known to seek out areas where salt is deposited and lick at it. Salt is necessary to maintain ade-quate levels of serum sodium, and animals instinctively recognise its importance. A creature with a full stomach is clearly not responding to hunger when it continues to lick at a salt-pan. Vertebrates often consume quantities of kaolin in the form of clay. They do so if they have eaten foods that contain toxic compounds, for these compounds are adsorbed onto the surface of the kaolin particles. Macaws in South America graze on clay beds when they have eaten seeds which contain a toxic component. Many birds, when choosing plants from which to build their nests, select species of plant which contain com-pounds that will help control mites and ticks, which could otherwise irritate the nestlings. Birds select a suitable blend of stone fragments to provide a grinding agent in their crops, which also help supplement the calcium intake they need to produce egg-shells. Many birds rub dust into their feathers, an act known as 'anting' since it was once believed to remove ants. The South American coati, a small animal a little like a raccoon, rubs itself against the resinous stems of *Trattinnickia aspera* and picks up traces of odorous compounds which seem to act as an insect repellant. Other plants of the same family, the Burseraceae, have been used in traditional medicinal remedies by North American indigenous peoples. It is also believed that the consumption of grass by carnivorous pets (like cats and dogs) is an indication of some intestinal disturbance.

But it is in the world of apes and monkeys that the use of medicinal plants has most often been observed. The capuchin monkey, best known as an organ-grinder's monkey, is named from the dark cap of hair which surmounts its head. This is said to resemble the cowl, *cappuccio* in Italian, worn by Capuchin monks (an order of Franciscans). The little capuchin monkey is familiar with a range of medicinal plants which it can bring to its aid. Some of them, including the leaves of *Piper* sp. and fruits of *Virola surinamensis*, contain active alkaloids. Others, like *Eugenia nesiotica* and *Protium* sp., are rich in essential oils. Chimpanzees in particular have been shown to use many medicinal plants. They do not chew them as food, but often swallow them whole which allows healing compounds to escape digestion in the stomach. *Ficus exasperata* is a relative of the fig which is often chosen by chimpanzees. Leaves are collected early in the morning and swallowed without being chewed. At the University of California at Irvine, separation and analysis of the constituents revealed the presence of furanocoumarins and 5-methoxypsoralen, which are already used in human pharmaceuticals.

The most frequent explanation for such behaviour is that

animals graze on leaves and fruits without any real purpose. The phenomena are dismissed as one of those myriad actions which animals carry out in their mindless lives as robotic automatons. Let us look more closely at a single example of how chimpanzees use a plant for its medicinal benefits. Chimpanzees in the Mahale Mountains in Tanzania consume leaves of the shrub *Aspilia mossambicensis*. There are several intriguing aspects of their behaviour, relative to this shrub, which scientific research reveals to have important medicinal implications.

○ The leaves are selected with far more care than is taken when choosing leaves of a normal plant eaten for food. Only young leaves are picked. The chimpanzee eats *Aspilia* leaves at the rate of five per minute, whereas they eat food leaves at up to 40 per minute. Analysis of the leaves has revealed the presence of a range of important medical compounds. The principal component is thiarubrine A, a red-coloured antibiotic which has a strong action against parasitic worms.

○ Each leaf is rolled round the tongue, and is swallowed without being chewed. The chimpanzee swallows each rolled leaf much like a human swallowing a pill – and for similar reasons. The leaves taste bitter and unpleasant, and the chimpanzee avoids the repellent taste by swallowing the leaves intact. It has also been shown that the active ingredient thiarubrine A is inactivated by stomach acid, but survives if the leaves are swallowed without crushing.

○ The chimpanzees go to the *Aspilia* plant within an hour of rising in the morning. This means that the stomach is empty, which is the circumstance under which the thiarubrine A can act directly on any intestinal parasites. Food in the stomach would dilute the active ingredient.

○ Chimpanzees do not always eat *Aspilia* leaves, even where they are abundant in the area where the animals live. On days when they do need the leaves, they undertake a special journey to find them.

○ Interestingly, there is a sex disparity in the consumption of the leaves. Female chimpanzees eat them more often than the males.

Two species of roundworm, *Trychostrongylus colubriformis* and *Caenorhabditis elegans* (nematodes, see p. 254), were tested in the laboratories at Irvine against extracts of the leaves. The extracts proved to be highly effective agents against both species of parasite. Toxicity tests showed that an adult chimpanzee could safely consume as many as 3000 of the leaves. In practice, they rarely eat more than a hundred at a time, which is well within safety limits, but provides more than enough thiarubrine A to kill any parasitic worms. The swallowing of the leaves whole is a vital part of self-medication by the chimpanzees. It is easy to prove: entire leaves can be recovered from the animals' faeces, so they pass through the gut intact, the soluble constituents being absorbed by the body during digestion.

Tests have yet to be carried out on the drugs that other plants contain, though their effects seem familiar enough to the chimpanzees. They consume the important components, even when the taste is unpleasant. Chimpanzees can become ill with intestinal disorders. When this happens they lose their appetites, become constipated and slow to move, and are clearly unwell. They will then seek out the bitter leaf plant, *Veronia amygdalina*. Only the tender young stems are collected, and then the juice is sucked out. It is extremely bitter to the taste. Examination of faeces collected over a period of time showed that the chimpanzees harboured gut parasites, which were greatly reduced in number as a result of the self-medication. Analysis shows that the leaves of *Veronia* contain high concentrations of sesquiterpene lactones, drugs which would harm the chimpanzee. The pith within the stems contains very little of the sesquiterpene lactones, but is rich in steroidal glycosides, which are highly toxic to the parasites and do the chimpanzee no harm.

Only one medicinal plant is eaten whole, and that is *Rubia cordifolia*. Among the compounds in this plant are anthraquinones, which have long been used in folk medicine. It also contains a cyclic hexapeptide which has caught the attention of pharmacologists. It is currently under investigation by the US National Institutes of Health as a possible agent for the treatment of some cancers. We have long known that the indigenous peoples of the tropical forests are aware of treatments which may give us valuable drugs for use in modern medicine. Curare, the poison-dart paralysing agent of the Amazonian tribes, is now routinely used in virtually every major surgical operation in today's Western hospitals. Here is a further possibility – that observations of the traditional behaviour of sick animals may offer new treatments for human patients. Not only may we learn from tribes of people living in time-honoured ways, we may learn from the animals too.

3

The Secret Life of
Birds

The world of birds is ancient and scientifically fascinating. The scaly legs of our avian cousins, their strong beaks and their nest-building are reminders of their dinosaur ancestry. Birds show cognitive ability. Ravens have been observed to solve the problem of feeding from a piece of meat hanging below their perch on a string. By trial and error the raven can work out how to pull up the string and hold it with its claw, time after time, until it can reach the suspended morsel. Ravens play, too. A raven has been seen to slide down a slippery slope on an ice-bound hill, repeating the process time and time again as though doing it for pure enjoyment. Ravens have also been seen playing in an airstream; repeatedly flying into a wind blowing over a cliff edge, and tumbling over in the eddies of air.

Our forebears knew much of the behaviour of birds. Science has already learned much from birds by observing them. The drug reserpine was discovered in *Rauwolfia serpentina*, after birds were seen to fall from the skies in a stupor as they tried to fly away after grazing on the plant. Birds gave us cocaine, after being seen to feed addictively on seeds of the coca plant. Legend says that the stimulative properties of tobacco were first discovered after birds were seen to become excited after feeding on the flowers. The Huichol people of ancient Mexico have a folk-tale which tells how hunting birds feed on tobacco to gain stamina and strength to fly higher and higher into the sky. It is said that the ancient Greeks took to cannabis after seeing how wild finches which had fed on the seeds became docile, tame and content to be petted. The Chinese teach that rice wine was discovered when sparrows stored rice grains inside hollow bamboo stems. As winter approached, the secret stores of rice were flooded with rainwater and natural fermentation soon started. The strange behaviour of the birds led the local people to try the brew, and in this way birds gave rise to the origin of wine-making in China. It seems no more than a fanciful legend, until you realise that the modern ideograph for *samshu*, rice wine, consists of the ideographs for 'bird' and 'water' combined.

Birds seem to be unemotional and without feeling, yet we know that they conduct elaborate courting rituals, and display great passion and devotion to lifelong partners. Birds can inflict serious injuries on one another during territorial disputes, sometimes fighting to the death; yet show self-sacrificial behaviour in defence of their young (which indicates genuine altruism and parental devotion). Many birds pair for life, and pine at the loss of a partner. Senses in birds are highly developed. They have some senses more finely attuned than those of humans. Pet birds show grief and appetite loss when a loved one is lost. Parrots will pine indefinitely if isolated from their partners, will lose plumage and show no interest in life. Sometimes they cease all activity, literally giving up living, until they die. Birds are rational, thinking, caring, emotional creatures. They may seem otherwise to us, but that is because we look at them through human eyes. Looked at through the eyes of a bird, humans would seem no more than instinctive automatons, propelled by subconscious urges we cannot control and lacking in love of our fellows.

Song birds

Birdsong seems to be one of the purest sounds. Hearing the rounded, limpid call of birds perched high and serenading the world around is a glorious pleasure. The lilting cadence of sweet notes is a delight to the ear that sits in uneasy contrast to the birds' appearance as cold and emotionless creatures. They cock their heads and peer as though unmoved at their surroundings, yet may react to an intruder. If they are a bloodthirsty species (like the traditionally endearing British robin, which is perhaps the most violent garden bird of all) they may attack it with unbridled rage. None of this is betrayed in the birds' facial expressions, because they have no facial musculature, as do mammals. But even though they betray no emotion, inside their emotions churn. Birds will pine for a mate, grieve for their offspring or long to return home. If they had soft cheeks and pliant faces we would see them as the most responsive of creatures. They do react to life's vicissitudes. Humans can tell one bird from another by listening to its song, but birds can tell far more. They can distinguish individual birds from the songs they sing, and recognise strangers as soon as they open their beaks. Birds have strong regional accents. Outsiders are rarely tolerated where territorial boundaries have been set. The sense of sight in many birds is extraordinarily acute. Smell seems much less important, in terms of providing input on reality.

Song as a means of communication is widely used by birds, though their hearing is not very acute. Whereas mammalian ears have three ossicles to amplify the sound vibrations and transmit them to the auditory nerves, birds have a single bone, the columella. It does the job less accurately, and birds can hear only up to a frequency of about 10,000 Hz. The hearing of a young bird is like that of an old man. The use of sound by a bird depends on its ability to vocalise, of course, but it depends far more on the effect of the environment on the sound once it has been emitted. It is tempting to assume that woodland would have a degrading effect on the transmission of birdsong, whereas open spaces would assist the travel of sound over long distances. In practice it is not so simple. In a wood there are many factors that deaden the sound, the damping effect of the mass of foliage being the most important. However, temperatures are constant, the atmosphere is stilled and the trees can reflect sound. In the open air there are problems when reflected sound waves from the ground interfere with direct transmissions from a bird (these actually cancel out some of the sounds), and scattering effects are caused by layers of air at different temperatures.

Birds have adapted to these realities. The song of forest birds is mostly in the form of pure tones. Notes as low as 500 Hz, or as high as 3000 Hz, are quickly attenuated in forest conditions. By contrast, notes between 1500 and 2500 Hz travel twice as far before they are lost in the background. It is in this range that most woodland birds sing. Birds' ears are adapted, not only to hearing their own songs, but also to the sounds of the natural landscape. Sounds like a frog croaking can be transmitted directly upwards for 1 km (over ½ mile) or even more, and it is believed that birds use their sense of hearing to monitor what is happening on the ground beneath them.

What birds hear

Conventional sounds are not the only auditory input that birds can utilise. They may not be able to hear notes as high as we can, but they can hear notes that are lower. Birds can hear lower frequencies than humans can, down to 10 Hz or even 5 Hz (with some response even to 1.5 Hz, almost one wave per second).

What are they hearing? The distant sea-shore emits low frequencies. Wind can set up similar effects, as could an advancing weather front. Storms are also sources of these very low sounds. Higher frequencies become quickly attenuated with distance, whereas lower vibrations can travel farther. The lowest notes of all, these infrasound frequencies, can pass through the atmosphere for hundreds of kilometres without being lost. Maybe birds monitor far-distant messages from wind and rain, waterfall and sea-shore. They are listening to signals which our ears cannot recognise, as sounds, at all. Perhaps birds do not have a continual set of thresholds, but change with the seasons. There is some evidence that a goldfinch hears best around 2–3 kHz for most of the year, but at higher frequencies during the breeding season. In May the upper threshold rises, peaking at up to 5 kHz in August, and settling down again during the autumn. The goldfinch is thus able to respond to higher sounds during the period of greatest activity. It seems that the sex hormones regulate the way a bird hears, and there is an annual cycle which may be widespread in the bird world.

Studies of the internal structure of a bird's ear under the microscope suggest that birds may have a different mechanism for appreciating sounds. We tend to hear a wide range of sounds, most of them higher than the noises to which we pay most attention. Our voices, for instance, send out sound of a few hundred Hertz but there are resonances in such sounds which may reach 15 kHz. Listen to a chiming bell (like the sounding of the hour by Big Ben) and you can hear the interweaving resonances which seem to go on for a long period after the bell has been struck. The structure of a bird's ear suggests that they hear purer sounds. Humans have an amazing ability to distinguish a tiny change of frequency in a note. Some birds come close to that. A parakeet can identify changes of frequency of less than 1 per cent, which is better than a pigeon or a cat, and almost as good as a human. Birds have a far more finely tuned ability to separate sounds in time. Birds trained to distinguish a single click and two clicks together could distinguish clicks separated by as little as 2 milliseconds; humans need 30 milliseconds to identify two separate clicks. This may well be related to the complicated trills and beats set up when many birds sing. In order for us to study birdsong, we have to plot it out as a phonogram, or slow it down on a tape-recorder, before the complexity emerges. Birds, with their finer ability to separate events in time, can make these distinctions as they listen to one another. Birds also distinguish different amplitudes (loudness); in this respect they are about as good as we are at separating sounds on the basis of loudness.

Wisdom and the owl

Senses in some predatory birds are very highly developed. In the owls they are tuned to a remarkable degree. The barn owl *Tyto alba* can detect the slight rustling made by moving mice, and even in total darkness can sense direction with an accuracy of 1°. The ears of an owl can detect the different times at which sounds reach the two ears. This is often less than a ten-thousandth of a second. Rustling sounds are of high frequency, and owls have a range of hearing that reaches far above that of most song-birds. Recordings of the rustling made by mice show how owls can strike directly at a loudspeaker in total darkness. If the higher frequencies are cut down, the owl's response changes. When there are no frequencies higher than 8 kHz the owl's accuracy goes down from better than 1° to worse than 7°, and if the highest frequency is restricted to 5 kHz the owl listens with interest, but makes no attempt to strike at the target.

Their sense of hearing is intensified by the extraordinary construction of their external ears. The two ears are at different heights on either side of the head, and there are intricate skin folds around them which help to concentrate incoming sounds. The opening to each ear is covered with feathers through which sound can pass unimpeded, and behind these are parabolic reflectors made from very dense feathers which reflect all the sound towards the eardrum. If a tiny microphone is placed within the ear of a dead barn owl it is possible to find how the owl's hearing can be controlled by moving its head. At high frequencies above 8.5 kHz the owl's ears prove to be highly directional. Even a turn of just 1° either side causes the sound to become quieter. As the frequency of a note is raised, the sense of direction steadily improves. An owl with incoming sound perfectly balanced in both ears would find itself staring directly at the sound source.

Why do owls have differently placed external ears on each side of the head? It has been argued that this may be because their acute sense of hearing allows them to make use of the minute time delay between the arrival of sound at each ear. If that is true for some, it does not apply to barn owls. As mentioned above, they can be induced to respond to tones played back through loudspeakers, and barn owls react exactly the same to single notes as they do to a continuous tone. Owls can hear things we could not detect. One of the most impressive examples is the way that the great grey owl *Strix nebulosa* hunts for lemmings in winter. The owl flies down towards the snow and strikes through the unbroken surface to seize lemmings that are hidden from view. The owls hear the movements of the lemmings (perhaps their chewing on seeds) and stoop on them without needing to see their prey. If ever there was an example of an acute sense of hearing we could not emulate, this is it.

Navigation by echo

Bats and whales are among the creatures which can use echolocation. There are some two hundred and fifty species of bird which fly at night, yet only a very few have developed echo-location as an aid to navigation. Some cave-dwelling species use echolocation to find their way in the dark. Cave swiftlets of the genus *Collocalia* produce clicks as they fly inside their roosting caves. These are the birds from whose roosts bird's-nest soup is made. The flying birds emit high-pitched double clicks, a moderately loud initial burst of sound followed by a louder second click a few milliseconds later. They can produce these clicks at rates of up to 20 per second, and can locate structures in the dark as small as 6 cm (2½ in) across. *Collocalia* eat insects, but do not use echolocation to find them. The clicks start as the birds fly into their caves for the night.

They sometimes use their echolocation sense outside the cave, for these birds like to play with one another and make swooping flights over water during courtship, or practice dives towards one another when fully fed. The need for echolocation in the caves is well documented. For example, the species *C. esculenta* nests in partially lit areas of caves and, because they can see where they fly, they have not developed echolocation. The only other genus to have developed echolocation is the fruit-eating *Steatornis* of South America. This species, known popularly as the oil-bird, has large eyes which are adequate for vision in low light, but uses sonar by sending out sounds when flying into its home cave. It emits rapid bursts of clicks only 3–4 milliseconds apart. The bursts are emitted at the rate of 5 per second when the birds are flying down the open centre of their caves, increasing to a peak of 11–12 bursts per second as they approach the ledge on which they have chosen to alight.

If the birds have their ears temporarily blocked, they become disoriented and fly into the walls. As soon as their hearing is restored, they resume their dare-devil high-speed flying and can find their way around their crowded caves with unerring accuracy. Sometimes *Steatornis* have been heard to click while in the trees where they feed on fruit, so it may be that they use their night-location sense to help them find their way around in the day-time.

Birds and the sense of sight

Our capacity to see fine detail in colour and at a distance is matched by very few other creatures. Some birds, however, can see better than us. Birds' eyes are huge, occupying a very large proportion of the space in the skull. In some birds the globes of the two eyes touch in the mid-line of the brain. There are three different types of bird's eye:

o Flat eyes with a short axis from front to back, typical of ground-dwelling birds including pigeons and poultry.

o Globular, rounded eyeballs, typical in conventional perching birds and birds of prey which hunt in daylight.

o Long or tubular eyes, which are longer from front to back than their width from side to side, found in nocturnal birds of prey.

The anatomical adaptations in birds' eyes are remarkable. We have seen that the cornea is the main focusing component of the eye in animals which live on land, and in some birds there is a special structure (Crampton's muscle, named after its discoverer) which encircles the cornea and can increase its curvature. This offers an immediate means of making the eye accommodate far more efficiently than human eyes could manage. Another accessory not present in humans is Brucke's muscle, which presses against the edge of the lens. It can increase the power of the lens when necessary for close-up vision.

These extreme changes are necessary in birds, like the cormorant *Phalacrocorax*, which live above water but hunt for food beneath the surface. The quick change in the profile of the lens means that they can focus underwater as well as they can in the air. Not all diving birds have this facility. The brown pelican *Pelicanus occidentalis* spots its prey while it is flying, and then dives in to scoop up the fish with its expandable beak pouch. It closes its eyes as it hits the water and relies on momentum to bring it to its target. Birds of prey have the ability to see the finest detail. Most birds do not see the world as clearly as humans, but the best eyes in eagles exceed our ability to perceive fine detail. Large eagles can see details three times better than humans. Owls have very fast lenses in their eyes. Most pocket cameras have a lens rated at about f/3.5, while the human eye, as we have seen, is rated at f/2.1; an owl's lens is rated at f/1.3 which is faster than all but the best of camera lenses. Owls may not be able to sense finer details than us, but they can see far brighter images in dim light.

Birds can see in colour. The spectral range visible to a particular bird is related to the bird's lifestyle – nocturnal species do not perceive colour as well as most day-flying birds, but they can pack a million light receptors into a single square millimetre of the retina and so can see fine detail. The spectrum seen by a bird also depends on its ancestry. Many domestic fowl are descended from forest-dwelling wild species and still have eyesight which is corrected for the green light that predominates under trees. Within the structure of the retina of nocturnal birds is a layer which reflects rays of light entering from the front of the eye to give the retinal cells a second chance to capture fleeting images. The effect is found in other essentially nocturnal animals, like foxes and cats, which is why their eyes are so easily caught in the lights of a car at night.

Birds can perceive forms of light which we humans cannot distinguish. Many birds rely on polarised light, to which we are blind. Many of them specialise in the blue end of the visible spectrum, and can see images into the ultraviolet. The lenses and corneas of birds are transparent to ultraviolet, whereas the lens in humans (and many other vertebrates) is slightly yellowish and prevents the transmission of ultraviolet rays. Some human patients have had their natural lenses replaced with a plastic prosthesis, and report that they can see patterns on flowers where no patterns at all are visible to the conventional human eye. This operation increases our sensitivity to the far purple end of the visible spectrum, and it would be intriguing to document exactly what these post-operative patients can see which the rest of us cannot. These special senses of sight allow birds to navigate, polarised light in particular being an important aid to direction-finding under the open sky.

Compass navigation

Many birds are sensitive to the earth's magnetic field. Experiments have demonstrated the truth of this belief, and in recent years the microscope has shown us where the sense organ lies. Some of the earliest indications came from young pigeons which were fitted with a small bar-magnet on a collar. Their behaviour was compared with an equal number of young pigeons which had been fitted with a metal blank to act as a control. When taken away from their loft and released, the control pigeons returned, after feeding, without difficulty. The pigeons wearing bar magnets became disoriented. Clearly, their magnetic sense had been impaired and they had lost a crucial means of learning the way home. Similar experiments on experienced homing pigeons produced different results. With magnets fitted, homing pigeons could return to the loft without delay if the sun was shining. However, on a dull and overcast day, they tended to lose their way.

How important is sight to a pigeon? Would a sightless pigeon find its way home through the use of its other senses? The answer is 'yes', but the experiment to determine this was difficult to design. The primitive notion of experimentation with sightless birds would condemn them to death through starvation, for they need their eyes to find their food. Experimenters in America fitted pigeons with semi-opaque contact lenses. These prevented them from seeing any landscape details, though they could see close up. The lenses were made of gelatine, so that any pigeons which did not return to the loft would find that the lenses dissolved away harmlessly, restoring normal sight after a few days. The birds were taken from their loft, and half were fitted with semi-opaque lenses; the other half were given similar gelatine lenses which were clear, and did not affect vision. These were the controls, to see if any change of behaviour resulted from the mere presence of contact lenses in the eyes. Pigeons can fly home in a day from a release point 1000 km (about 600 miles) distant, and both groups of pigeons flew home without difficulty.

Clearly, the effectively sightless birds were not handicapped by this lack of a normally essential sense. It is apparent that these birds can use many cues, visual information and light from the sun being among them. When the day is dull they rely heavily on magnetic sensitivity and if this is blocked, the task of navigation becomes much harder. In another experiment the pigeons were fitted with an ultra-light magnetic coil and a battery. Half the animals had north at the top of the head, the other half had north pointing downwards at the nape of the neck. These tests produced an interesting result. On a sunny day, homing went on as normal. When the day was overcast, the pigeons with north pointing to the base of the neck headed homewards, whereas the others (north towards the top of the head) went off in almost the opposite direction.

Disturbances in bird navigation were recorded around the

Project Seafarer antenna of the US Navy, which produced a strong electromagnetic field and blocked out the background magnetic field. Similar problems have been noted when there is a solar storm, and the earth's magnetic field is temporarily disrupted. Observations of the European robin *Erithacus rubecula* show that, though they have an ability to navigate by the stars, the sense of magnetic direction can be used if the stars are not visible. There is something strange about a sense like this, which humans do not possess. If bacteria can navigate by the earth's magnetic field (see p. 268), it should not surprise us to learn that a similar sense is present in more highly evolved creatures. The problem has been to locate the sense organ, and to find how it works. Microscopical examination has shown that there are granules of ferric oxide, magnetite, buried deep in the heads of some birds. These respond to the earth's magnetic field, and signal to the brain of the bird how it is oriented as it moves from place to place.

The detailed functioning of these organs remains to be unravelled, but at least we can see a connection between the simplest of bacteria and a highly evolved homing pigeon. Do humans share this sense? Tests have been discouraging, for blindfolded volunteers seem to set off in random directions, unable to deduce anything about their orientation from any subliminal magnetic sense. Once again, we can see an important sense in other creatures which we ourselves seem to lack.

Birds and the human world

The geese that warned of a raid on Rome by the Gauls in 390 BC saved the city from occupation. They picked up the footsteps of the warriors advancing in the darkness and began to honk their warning. Capitoline geese, in memory of this event, are kept to this day in the cathedral cloisters in Barcelona. During the Second World War a quacking duck ran loudly through the streets of Freiburg, Germany, and aroused the inhabitants so that they had time to shelter before an allied bombing raid. Birds are responsive to infrasound and perhaps they were warned of an impending raid by a steadily increasing intensity of pressure waves from the approaching planes. These are warnings given through sensitivity, not prescience.

The ability of birds to find their way acquires a special romance when it concerns a rendezvous with a person. A young boy in Virginia kept a pet pigeon at his home. One night the boy fell ill, and was taken to a hospital many miles away. A little later, in the middle of a heavy snowfall, the pigeon was reportedly found tapping on the outside of the window near his hospital bed. In 1980 at Cape Cod a seagull, regularly fed by an elderly widow, conveyed a warning which saved her life. She fell from a cliff-path while walking. The gull, which had been circling overhead, flew to her home and was said to have pecked repeatedly at the window to attract the attention of her sister. Once she had been alerted by the gull, and had started to follow it towards the path, the gull flew ahead in swooping dives, landing from time to time and looking back as though watching to check that it was being followed. It brought rescue to its human patron, who remained convinced that the gull had saved her life.

Language of the gulls

Gulls are remarkable birds, somewhat larger than you would expect (for they are usually seen by people at a distance) and intensely sociable amongst their own kind. Gulls live in communities, travelling in flocks and constantly interacting with one another. Above all, they have a complex language. Specific calls are used to warn of danger, as a marker during the breeding season, and to attract others in the flock to food. Not all their communication is through sound, for they also use precise body movements to convey their feelings.

Gulls communicate with their young, even calling to them through the eggshell before they are hatched. There is a special mewing call used by the parent to her newly hatched chicks which alerts them to her arrival at the nest. Only when the young peck at the tip of her beak does she regurgitate a cropful of food for them. As they mature, the young gulls develop behavioural patterns through innate mechanisms which we presume to be genetically predetermined. That is the secret of the origin of the responses, for birds raised apart from their parents still develop the same repertoire of body movements and vocalisations. As we shall see (see p. 123), learning and teaching are also important and there are countless other examples to remind us of this fact. We have seen that the pecking of a young gull on the beak of the parent signals the delivery of partly digested food. This sounds like an automatic response, like the sound of a bell when the button is pressed. In reality it is much more complex. Birds use their brains. The parents take time to make sure that the young understand the signals, and if they are slow to start they will teach them what to do. The baby gulls take time to learn how to peck accurately. That can be demonstrated by presenting the young chicks with a model target in the form of the head of an adult. A film camera records the scene, and each peck is recorded as a mark on a photograph of the head. A newly hatched chick records pecks in a rounded pattern, some of which strike the beak, but many of which miss altogether. Two days later a repeat performance has the young bird aiming far better. Every peck lands on the beak. The 48-hour period of learning by experience, and the increasing steadiness of the chick, have produced their effect.

Each chick learns to peck with optimum force. If it misses the target, it is likely to be pecking with extra vigour by the time it eventually makes contact. If this happens, the rebound force of the impact can throw the chick backwards. Within a day or two it learns not only to aim straight, but to peck with the correct force. A young gull chick in the nest greets the returning parent with interest. It turns to face the parent and will peck at the proffered beak. This is the signal for the parent to regurgitate food and, after a moment or two, some partly digested fish will be produced from the parent's crop and set down beside the chick. What if the chick does not start to peck? It may fail to recognise the stimulus, or have forgotten what to do. If this happens, the parent maintains its posture, with its beak close to the chick and waving slowly to and fro. It is exhibiting its beak to the youngster and, by moving it, making it more attractive. The youngster is encouraged to peck at the moving object. Once it has performed its peck, showing the parent that the lesson has been learned, it is fed.

Sometimes the regurgitated food is ignored by the young-sters. If that happens, the parent will lower their beak towards the food, pointing to it, drawing the chick's attention to it, encouraging it to feed. In most cases an unwilling chick, or one which is slow to learn, will then start to feed. The parent watches intently, to ensure that the chick is feeding properly. What if the food is still ignored? The adult will pick up some of the food and hold it in front of the chick's beak. As the chick feebly pecks towards the regurgitated mass, the parent moves the food towards the beak of the chick to encourage it to feed.

Only when the parent is satisfied that all its chicks know what to do will this routine of caring and training come to an end. As the chicks feed, the adult gull monitors what is going on, and maintains a watchful eye for predators or strangers which might invade home territory. At the end of the feeding session, the youngsters are satiated and no further food is eaten. The parent bends down to pick up all remaining scraps, and consumes them all. The nest is left clear, the young are properly taught how to feed, and only then do the adult gulls continue to feed themselves. They take raising their youngsters seriously.

In mature birds, certain bodily attitudes and movements are clearly signals of submission, aggression or nest recognition. These gestures are universal throughout a given species. In different species of gull the gestures are recognisably different from one another, though they are still somewhat similar. It is this spectrum of signs and signals that encourages us to assume that all 35 species of gulls are descended from a common ancestor, the behavioural signals having become modified as the morphology has changed. Just as morphologists and bird-watchers distinguish species by their external markers, it is possible to distinguish gulls by their visible movements. The movements and signals of the large gulls belong to a common category, just as do those shown by the hooded gulls, and they are in turn recognisably different from those of the kittiwake and its related species. Classification by body language parallels classification by genetics or morphology.

What I find particularly interesting is the essential universality of these postures. An upright stance, beak poised, is an aggressive gesture to a rival bird. It also looks aggressive to the human observer. When gulls fight, they peck wildly at each other and pull out feathers if they can, which they toss to one side as they return to the attack. Gulls engaged in a territorial dispute on dry land use a 'redirected attack' in which they perform exactly the same movements – but directed against the grass growing between the rocks, rather than their enemy. In this form of behaviour an angry-looking gull pecks at the vegetation, pulling fragments out of the ground, and tosses them to one side in a frenzied attack on an inert target. This makes the point, without incurring the risk of physical harm to either participant. In this instance too, it is quite clear to the human observer what is going on. The gull is clearly angry; its stance is obviously aggressive. There are comparisons with the way an exasperated man might bang his fist on the table, kick the furniture or tear out his hair (do people really tear their hair?) and we do not need training to recognise the signs.

When a gull decides to terminate the display of anger, it is done with a sideways flick of the head. Commentators have said that this movement is identical to the way a bird flicks some sticky material away from the beak during feeding, and conclude that this must be the origin of the movement. That is a mechanistic interpretation as seductive (and as misleading) as to assume that two human gestures have a common origin simply because they look the same. Just because we may blink our eyes if responding coquettishly to an amorous advance, exactly as if they are irritated by smoke, does not mean to say that we learned the gesture in one situation and applied it elsewhere. I can imagine our alien observer reporting back to base: 'Ah! The human subject blinks at cigarette smoke just as she blinked at her lover. Clearly she loves cigarettes!' The dismissive flick of the head at the end of an exhibition of attacking behaviour is recognisable, even to us humans, as a gesture which terminates the display. In so many ways we can see in other creatures a litany of responses that compares with our own.

Am I moving towards anthropomorphism? Quite the contrary. This can be no attempt to force a human-centred interpretation on a bird's behaviour – they were here first. As I sit in the middle of the gull-crowded cliffs on the island of Steep Holme in the Bristol Channel, all I am aware of is the extent to which I see gulls conveying their feelings to the human observer. It is the universality of emotions one witnesses, not something arbitrarily applied to a creature which doesn't warrant it. This is no projection of human sensibilities onto animals, but a realisation that they have always had gestures of their own. Birds pre-date such impressionable newcomers as *Homo sapiens* by millions of years.

Observers categorise the body language of gulls into eight manoeuvres:

○ *Rest.* The gull stands still, body somewhat slouched, and with the head slightly lowered.

○ *Hunched.* This is similar to the posture at rest, but the head is drawn down into the shoulders and the beak is slightly raised.

○ *Oblique.* Here the bird is reaching out and calling over a distance.

○ *Facing away.* The head is turned, or flicked, to one side.

○ *Mew.* Soft sounds are made, as opposed to the long-distance cry, and the gull hops from one foot to the other.

○ *Choking.* As the head moves rhythmically up and down the gull describes jerking movements with its beak.

○ *Forward.* In this characteristic position the gull stands high, but bends the fore-part of the body lower than the rear, while raising its head.

○ *Upright.* The gull stands boldly erect, neck extended, the head held high and looking slightly downwards

The choking movement is one used by all species of gull, though some use it to mean something different. Herring gulls use it as a means of bonding, for males and females do this together when they stand at the nesting site. Male kittiwakes use much the same movement to advertise the availability of a nesting site to interested females, and to warn away rival males. The movement is like that made by the bird to balance itself before squatting down on the nest. Perhaps that's where it originated. Today it has a complex significance, and it is as though the birds are communicating.

When the gull stands erect and adopts the *upright* pose it is making a threat. It stands tall, eyeing the opposition, and is clearly ready to strike if the need arises. Sometimes the male gull will raise his feathers slightly, as though about to open his wings. This makes the bird look larger and more intimidating, and is also the action immediately prior to lunging forward (the wings act as counterbalances for the forward thrust of striking an opponent). This often deters the rival bird from further display, though in rare cases a physical fight follows.

If the threat is not successful, and the opponent does not draw back but stands firm, then the gull may raise his beak and allow his feathers to settle back on the body. It is a gesture of submission, or at least it shows that he concedes the right of the other bird to be there. An alternative gesture is to turn broadside-on to the opponent, which is taken to be a gesture of fear. Eventually, one of the gulls accepts defeat and lowers his head, or even scampers away. In most cases of confrontation, a physical fight is avoided. This is true of most animal species, and is even true of ourselves.

The adaptation of the different signals among the many species of gull reflects their particular habits. The kittiwakes, for example, spend much of their time feeding on the open sea. When they nest they do not choose open land, as do typical gulls, but become cliff-dwellers. On the cliff-face they are reasonably safe from casual predation, and territorial disputes do not arise. As a result, they are rarely heard to utter the distress cry, which is a common sound from ground-nesting gull colonies. But they do show a sign known as *head-flagging* from an early age. A threatened bird averts its gaze and looks away, as a sign of withdrawal from conflict. Interestingly, the large gulls sometimes do this too; it is one of the expressions to which they sometimes revert as adults when a conflict looms. What is special about kittiwakes is that they acquire this expression from a much earlier age. When there is a 'universal expression', then, its development can be accelerated if the demands of a specific lifestyle require it.

The larger gulls defend their territory with vigour. They spend much time acquiring a stretch of ground and mark out a site they henceforth regard as their own. Bigger birds claim a greater area, though the amount of space is not absolute. It depends on the degree of crowding of the colony, and adult males in over-crowded areas are content to claim less ground space for themselves. If a marauding bird approaches the nesting site of one of these gulls, the occupant responds with an initial call. It is a first warning. Should the intruder continue to approach, the gull will walk towards him, watching carefully as he approaches. He draws himself up taller as the distance decreases, until he adopts the classical *upright* posture. It is likely that the intruder will retreat at this point. If he does not, the nesting gull may well start a *choking* routine, as though communicating to the other bird. The message is clear: he is showing that he is in occupation, and is ready for action. As a rule, it is at this point that even a determined intruder will turn away and retreat.

Gull species which nest on inaccessible cliffs respond differently. They have only a nest, and no extended territory to defend. In consequence, they rise to their feet as an intruder flies towards them and – if the approach continues – start bobbing up and down in a *choking* manoeuvre. All gulls use this movement when they are on the nest, and it has been suggested that the movement is similar to the gestures by which new material is added to a nest. There is no reason why not. In our own language a word – 'home' for example – is incorporated into other expressions, including 'home-body', 'home-maker', 'home-wrecker', 'home-lover'. The alien scientist watching us at work would soon see the similarity of the roots in each term, and there's no reason why we should not find similar derivations for the roots of sign languages in species that use movement instead of speech.

Towards a dictionary

How far can we relate these episodes to what the gestures mean? It is quite clear what the gull meant in the nest-defending postures I have described. He passed from 'Oh-oh, who's that?' to 'Do you mind? Someone is nesting here already' to 'I'm warning you, keep your distance!' and finally 'Right, that's it!' As the would-be aggressor retreats and the bird settles back into its nest, we could even have the gull's equivalent of 'Kids today! I blame the parents.' Such translations are not so far from the truth. I say this for two important reasons. Firstly, the substitution of our meaning for the meaning in a gull's gesture is no different from translating one language into another. Translated terms do not necessarily mean what they say. Thus, we customarily translate the English 'goodbye' into the French *au revoir*, but it is an incorrect substitution. The nearest we can get to the French for 'goodbye' (which means 'God be with ye') is *adieu* (literally '[go] to God'). *Au revoir* means 'To the next time we see each other'. Its best rendition in English would be the almost literal translation 'See you later'. When we translate from one language into another, it is our intention to substitute not identical words, but closely similar meanings. It is realistic to substitute grammatical conventions of one language for colloquial expressions in another, and we can apply these principles to the interpretation of animal sign language. Indeed, were we to write down the literal translations of these gestures we might find it easier to systematise them. Some scholars use voice-prints of birdsong, because that makes it easier for humans to study, and in my view there's not the least reason why we should not eventually compile a Gull–English Dictionary as an aid to comprehension.

Love-birds

Green, gold and orange, sometimes suffused with a hint of blue, the little parakeets of the genus *Agapornis* have long been known as love-birds. There are two reasons: they show intimate and endearing courtship behaviour, and they pair off for life. In both respects they seem to have much in common with idealised human relationships. In many parakeet groups the males and females of each species have different plumage, as they ordinarily do in the great bird kingdom, whereas in *Agapornis* the two genders look the same. Little wonder that courtship rituals are stylised and elaborate, for without that it would be hard to tell one sex from the other.

The pairing of love-birds starts with the selection of compatible partners. Single birds meet in groups and inspect each other. They approach, each holding their head towards the other, sometimes trying to preen or groom each other. There are clear attempts to engage the interest, and perhaps the approval, of a partner. Love-birds take their time to choose a mate, and do not fall too easily for the first potential partner to come along. The male of a pair makes frequent attempts to woo his partner. In all the species of love-bird it is common for the female to rebuff many of these attempts. This may be by indifference, or even aggressive rejection. Male love-birds at the receiving end respond in different ways. They start by sidling up to the female while turning from side to side on the perch. This approach, known as switch-sidling, is the first move in courtship. Following this, the male sings to his partner in a twittering song pattern which is complicated and inconsistent. It is clearly some form of language which they share. Her most likely response is to turn her back and disappear into the nest cavity.

At this point, males often stop where they are and scratch their heads. It looks, to a human observer, like a gesture of frustration and the behavioural situation in which the bird finds himself substantiates that this is exactly what it is. The foot with which he scratches his head is always the one nearest the mate. He has moved his claw, ready to mount his betrothed, and when he is rebuffed he uses the raised foot instead to scratch the top of his head. Human parallels are too obvious to cite. I have to stress that this is not anthropomorphism. Indeed, it is an anthropocentric culture that has encouraged generations of scientists to decry these comparisons. If anyone is 'copying' anybody else, it is we who are copying the love-birds.

The pairing of the birds brings much mutual interaction. As they meet after a separation, they perch together and indulge in bobbing movements of the head. Up and down they bob, watching each other as they do so. Once the union has been re-established, other cooperative activities can start. The exchange of regurgitated food is common. In all species, males give food to their mates. In a few, females also feed their male partners. A female love-bird signals she is ready for copulation by fluffing up the plumage on her head. She leans forward eagerly, raising her head and tail to the obvious pleasure of the male.

Complex behavioural conventions underpin this activity. Not all of it is instinctive, however. Newly paired couples are awkward with each other and the males often incur the displeasure of their mates. It takes time for them to adapt to each other's preferences. Once a few broods have been reared, the two will have settled into an optimum pattern of behaviour, and rejection and frustration signals happen far less often. The males rarely twitter; at this stage they hardly ever suffer a rejection that leads them to sit and scratch their heads. The females seem ready for copulation more often than they were, and there are fewer signs of aggression.

Switch-sidling is the male's seductive approach to his mate, and as time goes by the time taken up by this activity is gradually shortened. Foreplay is diminished, for the females seem increasingly ready to copulate without the overtures. Conflicts between unrelated individuals take a ritualised form. The aggrieved love-bird strides purposefully towards the foe, the deliberate strides indicating aggressive assertiveness. A fluffing-up of the feathers and aversion of the head indicate submission and withdrawal by one of the two birds.

This complex ritual of communication between two adults means that there are few physical encounters. Some of the species indulge in play-fights when they demonstrate their speed and abilities. The two birds lunge at each other with their bills, almost as if fencing, and sometimes take nips at each other's legs. No other part of the body is ever bitten during these play-fights. Young birds in the colonies practise these manoeuvres with each other and are taught how to do it by older, more experienced birds. Although the response is innate, the skill comes with practice and instruction. Play-fights ordinarily resolve disputes. If matters are not resolved and a fight ensues, the birds angrily attack each other and can inflict serious wounds.

Love-birds are good parents. They construct proper nests (many birds of the parrot family lay their eggs in a cavity and do not trouble to construct a nest). If a love-bird on her first brood is given hatchlings of another species, she will raise them properly as her own, but from then on will refuse to have anything to do with her own chicks from future clutches if they are a different colour. A unique method of nest-building is used by these birds. They find suitable material (leaves or discarded paper) and divide it into small portions by using their beaks like an office paper-punch. These fragments are then tucked in among the feathers for transportation back to the nesting-site. It is an ingenious method, for most birds use their beaks to transport nesting materials. Is the facility inbred, or learned? There was one way to find out. The peach-faced love-bird, *Agapornis roseicollis*, carries its nesting materials tucked between the feathers, while Fischer's love-bird, *A. personata*, carries them in its beak. Both species are closely related and will hybridise in captivity. In the wild they are separated by at least 1000 km (about 600 miles) and accidental cross-breeding is impossible.

The hybrids hatched like normal love-birds. As the first season approached they began to establish nesting sites, and then came the time to gather materials. What would they do? One parent inherits the habit of carrying materials in between the feathers, while the other carries material in its beak. The conflict of instincts we might expect did, indeed, take place. The young birds seemed totally confused. They tried to tuck portions between the feathers, but they fell out; they tried to carry them in the bill, but seemed frustrated that they could not carry enough.

In the end, they transported small amounts of building materials back to the nest site in their beaks, but only after trying unsuccessfully to push some of it between their feathers. At first, only 6 per cent of the material they cut was carried back to the nest in their beaks (the peach-faced love-bird, which carries materials tucked in between the feathers, also carries about 3 per cent in its beak). After two months of learning, the hybrids were carrying over 40 per cent in their bills, though each flight began with a symbolic attempt to tuck it between the rear feathers, which they fluffed up just as the peach-faced love-birds always do. Two years later they were all carrying material in their beaks, but were still making a few symbolic movements as though they were trying to tuck material among their feathers. By the age of three years, the efficiency with which the hybrids carried material in their beaks was comparable to that of Fischer's love-bird, which normally does this. They had learned how to change their behaviour in the light of experience.

There is one intriguing postscript. By the end of three years, they had largely abandoned the need to make a symbolic attempt at tucking building materials in between their feathers, itself a special aptitude no other birds possess. Only once in a while did they remember that behaviour pattern and give it a try. On the few occasions that they did, they made a far better job of it than they used to. Over the years, they had improved their technique so that they were getting better at tucking material between their feathers. For these hybrids the method had never worked, but practice over the years improved their skills none the less.

In so many ways there are gestures, movements and behaviour patterns in these creatures comparable to our own. To pretend that love-birds acts like humans would be anthropomorphism. What, then, should we make of the fact that humans now act as love-birds have been doing since before we appeared? According to current conventions we should all be guilty of subconscious 'avemorphism' of a peculiarly invasive kind.

Imprinting and the duck

There is something familiar about ducks. Few tales are told about herring gulls or kittiwakes, and there are not many fictional stories about parrots (strange, since they talk). Where tales like *Jonathan Livingstone Seagull* are told, they record the ways of the creatures rather than anthropomorphising them unduly. Ducks are different. Stuffed ducks are found in most Western nurseries as playthings for children. Daffy Duck and

Donald Duck are just two of the many characters with which children are familiar. Edward Lear (1812–88), who was one of the finest bird artists ever known until his eyesight failed, wrote *The Duck and the Kangaroo*. Rudyard Kipling (1865–1936) had a duck laughing and talking in *The Brushwood Boy*, and of course there was Henrik Johan Ibsen's *The Wild Duck*. Our familiarity with ducks is extensive.

Ducks are a disparate group, covering several genera. Thus, the blue-winged teal is *Anas discors*, the wood duck *Aix sponsa*, and the spectacled eider duck is *Somateria fischeri*. The large muscovy duck, *Cairina moschata*, is an unusual member of the family and has some goose-like qualities. We cannot interpret their vocalisations, but there are clear body expressions they use to communicate during courtship. The familiar surface-feeding ducks, like the mallard *Anas platyrhynchos*, have a basic common language of ten expressions, which varies from one species to another. The ten basic expressions used by male ducks in courtship are:

1. Initial bill-shake.
2. Head-flick.
3. Tail-shake.
4. Grunt-whistle.
5. Head-up, tail-up.
6. Turning to female.
7. Nod-swimming.
8. Turning back of head.
9. Bridling.
10. Down-up.

These expressions are sometimes used in sequence, words becoming phrases; in all the species 4, 3 and 5, 6 are used in that fixed order.

Konrad Lorenz (1903–89) showed how the inciting behaviour common in ducks may have arisen. If a pair of sheldrakes are accosted by another pair who approach their territory, the female is the first to take action. She runs at them threateningly, hoping to frighten them off. As the distance decreases, the intruders, by their very presence, become more of a threat and if they do not move off, the duck will stop in her tracks and exhibit the reaction of fear. This turns the tables. She ducks down her head, turns tail and flees back towards her mate. Before she reaches him, emboldened by the greater distance separating her from the foe, she may stop and turn back to look threateningly at the other ducks over her shoulder. This is deliberate incitement to trouble. She is inciting them to assert their claim, and encouraging her mate to respond aggressively. Lorenz explained that this is because of her overwhelming instinct to keep an eye on the enemy. Mallards do it too, but in somewhat different circumstances. In this species, the duck may keep looking backwards over her shoulder when approaching the intruder. This means that she is forced to take her eyes off the enemy for a moment or two. Lorenz said that the backward glance has become such a ritualised part of the ducks' behaviour that they do it from time to time, even if it results in losing sight of the foe.

What is particularly interesting is the extent to which we, as unrelated to ducks as one could imagine, still understand the expressions. The aversion of the gaze, the toss of the head, are part and parcel of the way we interact when threatened. When we recognise an expression as coding for 'fear' or 'aggression' we are not anthropomorphising the antics of a bird. Rather, we are showing that we, too, know what they mean. This forms the universal language so many creatures understand. We know when we are 'threatened' by an enemy, whether it is a mugger on a dark street or a rattle-snake in the desert. The gestures of a dog or a male swan protecting its nest make its intentions as plain as would another human.

Learning bird languages

These examples show us that birds are more than mindless robots. They learn. They teach. They communicate. The purpose of the body movements of birds is to transmit messages, and each movement is an expression of its own. Like other expressions, they can be subtly modified in meaning, and can be joined together in a syntactical sequence to convey meanings of greater complexity. That is why I believe it is time for us to assemble dictionaries of Animal–English for all such observations.

Bird vocalisations exist for two purposes. The first is to arouse emotional responses, through wooing or aggression; the second is to communicate items of information, such as a warning or a territorial imperative. This simple categorisation reminds us of the parallels between our speech and the way birds vocalise. It is clear that there is more in common between us than there is in disparity. The basics of birdsong are inborn, but the rearing of baby birds in isolation from their kind shows how important is the learning component. A chaffinch reared away from the sound of birdsong develops an ability to sing, but the range is greatly restricted. The normal adult song lasts two or three seconds, and that is also the length of the song uttered by the isolated bird. Where it differs is that the song lacks any embellishments.

The familiar song of a chaffinch consists of two parts, the second of which ends with a tuneful flourish. In captive birds there is a single phrase, not two, and the colourful ending is absent altogether. What we hear is a basic song, lacking in fullness and of simpler form. When the bird is about a year old there is a critical six-week period during which it learns its song pattern. Once learned, it is fixed for life. A bird raised in isolation will soon learn a full song by listening to an experienced adult bird. It requires this learning experience to establish how to sing properly. What happens when several of these isolated birds are then raised together? They have a chance to interact, but not to learn the conventional song of properly taught adults. The results are fascinating. Given the stimulus of a mixed group, each of the young chaffinches immediately starts to learn. As the weeks go by their songs change, become more elaborate and rich in embellishments. They do not start to sing like normal adult chaffinches, but they lose their native, basic song pattern. The stimulus of interaction encourages them to invent new conventions, and to construct novel features in their songs. They perfect a new song language of their own.

Experience as teacher

In the years after the First World War, milk bottles with sealed tops became widespread all over Britain. Not only were they reasonably secure, even if a bottle was knocked over or tipped upside-down, but they were hygienically closed until the purchaser had collected the bottle from the doorstep and taken it into the kitchen. One day in 1928 in the sea-port of Portsmouth all that changed. One blue-tit worked out how to get at the milk. Perching on the edge of the sealed lid, it would peck into the lid to provide some purchase with which it could hold tight, and it would then seize the lid and pull it open. Other tits learned the trick, and within twenty years the phenomenon spread right across Britain. From then on, milk bottles were no longer always sealed when they were fetched from the doorstep.

Herons sometimes use bait to lure fish to their favourite sites. These majestic birds have been observed to scavenge portions of bread and carry them to their fishing territory. The herons drop the bread and wait until fish are attracted to it; they then catch and eat the fish. Herons do not eat bread, and bread does not occur in nature. However, they have learned the association between fish and the bait, and how to harness the association for their own purposes. If you watch a young seed-eating bird attacking a newly opened sunflower head, you can time it as it opens each seed. The first attempts take a long while, as the bird experiments with angles of attack and different methods of holding the seed secure. The second seed takes less time, and within a few attempts the bird has refined the technique to a high degree. From then on, opening a sunflower seed is far easier. The bird has the capacity to adapt its techniques to the situation facing it, and – most important – to experiment with a range of options and to learn from the best of them. These creatures are not bringing into force a pre-ordained and automatic reflex action. They are using their innate abilities, altering them in the light of experience and memorising the best option. Animals can experiment, and they can learn.

This reminds us of the importance of learning, and how the stimulus of communication acts to propagate spontaneous learning even in the absence of a teacher. Such creatures must learn. If the adults do not teach them, they will learn from one another. Learning is an imperative for many forms of animal life, and there are lessons we may learn from the world of wild animals which serve to remind us of our duties to our own young.

Inborn or learned?

Watch a goose caring for her eggs. She turns them regularly, and settles on them to maintain the perfect temperature they need for the young to develop. Every now and then she moves them around with her beak, keeping them turned and maintaining order in the nest. Should an egg roll away from the nest, she carries out a familiar manoeuvre: reaching over the egg, she hooks it back towards the nest with her beak and tucks it back in place. It is easy to interpret her subsequent look as one of satisfaction at a job intelligently carried out. The anthropomorphic interpretation is a trap, as you can see by presenting her with another type of egg. She will retrieve that, too. She will also retrieve a golf-ball, or a champagne cork. One report even said that a goose reached out for a square wooden building-block from a child's construction kit, so the object doesn't have to be rounded. There is no 'intelligence' here, for the bird is manifesting an automatic and inborn reflex action necessary for her species to survive. She is not recognising her own eggs at all.

Clearly many of the basic behavioural patterns of birds are inbred. Circumstances modify them, and so does the subsequent development of a species. How far can the instinctive patterns of behaviour be altered from outside? The collared dove *Streptopelia* provides some insight, for these birds (smaller versions of the domestic pigeon) have a courtship ritual which is well known and which can be modified for study. The male and female birds in this species are indistinguishable. For a veterinarian to tell one sex from the other takes an anatomical examination. The collared dove has a pairing sequence which is predictable. The male walks around the female, looking at her and cooing. If they pair off, they start selecting a nesting site and within the first day they will have decided on the position and started to collect nesting materials. The two birds share this task equally. Making the nest takes about a week, during which time the birds copulate. By this time the female has become strongly bonded to her nest. If an attempt is made to lift her free she will hold on so tightly that she can bring it with her.

By days 7–11 she lays her first egg, usually around 5 pm. The second egg follows two days later, around 9 am. Once the second egg is laid, the male takes a turn sitting on the nest. From then on, the female sits for about 18 hours each day, the male taking her place for about six hours. The eggs hatch in about 14 days. At first the young (squabs) are fed on a secretion from the lining of the crop, traditionally known as crop-milk. When they are 10–12 days old they leave the nest but continue to demand feeding from their parents. Over the following week or so, the parents indicate how the young should feed from seed they collect from the ground, and gradually offer less food themselves. The young pigeons are self-sufficient by the time they are about three weeks old. By then the parents are beginning to restart the cycle. The male is cooing to the female and strutting round her, bowing as he does so, and shortly afterwards nest-building begins anew. The cycle takes six or seven weeks to complete, and in idealised circumstances it can be continuous. The devotion of the parents to the care of their nest, and the rearing of their young, is obvious. However, they need to do these things in sequence. Thus a female or a male without their partner has no incentive to build a nest, or to care for eggs supplied to them. If a pair are put into a cage with a ready-made nest and eggs already in position, they ignore them completely and set about building a nest of their own (usually on top of the nest already provided).

If the eggs are re-positioned on top of the new nest each day, the adult birds start to sit on them 5–7 days after their first meeting. They obviously need this time of acquaintance and interaction to become ready to incubate the eggs. Perhaps they didn't like the situation, or were reacting to being handled? One way to test this was to separate the handling from the incubation period, so pairs of birds were kept in divided cages where they had access to nesting materials, but not to each other. After a week, the separating partition was removed. The birds could now start to interact, but here too it was 5–7 days before they were ready to start caring for the eggs.

Perhaps the birds simply didn't like sitting on a ready-made nest with its complement of eggs. To test this hypothesis, experimenters raised collared doves in a cage and observed them pair off, become a couple, build a nest and lay eggs. At this stage the nest and eggs were replaced with new ones from outside. The doves knew something had happened, for they did not take to the nest immediately. But after two hours of exploration and activity as a pair, they were sitting on the new nest as though nothing had happened.

What happens if the birds are kept in bare cages for the period during which they would normally be nesting and laying eggs? Pairs were kept for a week in cages fully supplied with everything except nesting materials. Normally by this time they would have constructed a nest over five or six days and started laying. If the birds were given a nesting site and a supply of materials on day seven, they indulged in furious nest-building, as though trying to make up for lost time, and then laid their eggs a day or two later. In short, doves do not necessarily nest just because they have a supply of materials. They need a period of time to interact and become acquainted with each other. If they have a supply of materials after that period of time, nest-building will start at once. The doves will not sit on eggs if they are presented with them in a ready-made nest, but will do so if a period of nest-building has occurred meanwhile.

We can see their behaviour developing through set stages. They need to pass through the pair-bonding stage before nest-building can be attempted. Thus, interaction changes their priorities: from being interested in each other, they change to focus on nest-building. This activity further changes their relationship, for it is a necessary prerequisite to the next change, from home-builders to parents. Later experiments showed that (for both sexes) the readiness to accept parenthood is related to the length of time spent together, and the strength of the response is increased if nest-building materials are available.

One classical experiment, done before the Second World War at London Zoo, showed that a female pigeon would willingly lay eggs and incubate them if she was in the presence of the male, but they were separated by a glass plate and could not actually meet. The sight and sounds of the male were enough to trigger her responses.

It is the presence of a mate which makes a female dove ovulate. The behaviour of the male is the key factor. If a female dove is in sight of the male, but separated by glass, the male will perform his conventional bowing and strutting display and the female will ovulate shortly afterwards. In one experiment, the males were castrated. When they were placed in the partitioned cages they did not start the typical male display. Only a few of the females ovulated when exposed to these emasculated male pigeons. Maleness, clearly, is the vital cue.

Caring for the young

Birds show great devotion to their young. A ground-nesting plover takes risks to protect her brood. If a fox approaches she crouches low over the nest, protecting the young. If the predator approaches too closely, then she scuttles away from the nest towards the fox in order to attract its attention. Her skittering through the grass mimics the action of a mouse or a vole, and the fox takes off after her. She resists any urge to take to the air, but manages to keep ahead of the fox as he tracks her away from the unprotected nest. Once she has led him safely away from the site of her nest, she takes to the air and flies slowly away, luring the fox onwards. Finally, she takes a circuitous route back to her young, leaving the fox bewildered, at a safe distance and facing the wrong way.

Many birds vocalise at their young – they talk to them. Often this starts before hatching. A mallard incubating her clutch of eggs will mutter sounds to the eggs. The tiny ducklings within the eggs will respond as they near the hatching time, and will utter sounds of their own. There is clearly some kind of language here, and the ducklings often initiate the exchange. Sounds have been heard coming from eggs without there having been any stimulation from outside, though the mother duck usually responds to such a message with calls of her own.

We do not know what they communicate, though there have been attempts to examine sound spectrographs or voice-prints from the sounds. It has been shown that the vocal communication peaks immediately after the hatching of the chick. There is a possibility that the sounds serve to tell the youngsters when to hatch. If eggs of a clutch are incubated, the hatchlings emerge over a period of a couple of days. A clutch of the same eggs reared by the mother will all hatch within six or eight hours, even though the chicks are of differing degrees of maturity. It may be that it is the vocalisation which coordinates this, for the chicks have to initiate the hatching processes by chipping their way out of the shell.

There are obvious benefits. A brood of ducklings hatching together allows the mother to rear them as a batch and to make sure she protects them all at once. If they were to hatch at different times, she would be neglecting one batch (eggs or hatchlings) while she cared for the other. The young get to know the mother's voice during this time, and this helps them to bond with her and to follow her when they are in the open. That is vital for the safety of a young and vulnerable duckling. It has been known for over half a century that these young birds bond to the first thing they see. Ordinarily it would be their mother, but if it is a human they see first they will bond to that person instead. They will bond to anything – even a scrubbing-brush or a mask. Since these experiments by Konrad Lorenz there have been many subsequent enquiries into this phenomenon of imprinting. It should be emphasised that in Lorenz's original papers the term was *Prägung*, meaning 'stamping' or 'embossing' (as in the production of medals or coinage). The translation 'imprinting' is not quite right.

There is an obvious developmental benefit in being attracted to your mother. This must be what lies behind the nursery-rhyme about Mary and her little lamb, which followed her everywhere she went. The lamb had been imprinted with Mary. In reality, 'imprinting' in the sense of *Prägung* is the wrong expression, for it implies permanence. In fact, birds learn to adapt to alternative objects or individuals quite quickly. In one experiment, a group of mallard ducklings were left in visual contact with a research worker for a continuous 20 hours after hatching. They became imprinted, and followed him devotedly. However, this was not the end of the experiment. In this instance the young birds were subsequently allowed to mingle with a mother mallard, who had herself hatched a clutch of eggs a few hours before. Within 90 minutes the imprinted duck-lings were all following the mother mallard and her brood, and it was impossible to distinguish between the imprinted birds and the ones she had hatched herself. They showed no further predilection for the research worker who had been their surro-gate parent.

Much of what we take for mechanical programming is a learning process. The imprinting of young ducks can be reversed. If imprinting phenomena are what we seek to find, then the experiments we construct are designed to demonstrate the desired effect and there is little interest in what happens later. A subsequent experiment might show that the mechanis-tic approach ought to be set in context with a more pliable and adaptive process of learning by experience, supplemented by unlearning through expediency. Young birds learn, and learn through their senses

4

Insects, Fish and Cold-Blooded Creatures

Cold-blooded creatures range from bugs and beetles to the octopus; from the lobster to tropical crocodiles. They have a range of senses and respond to the surroundings in a way that indicates a considerable problem-solving ability. Damage the nest or house of an insect and it will work out how to fix the damage, even though nothing like this will have happened to it before. Watch two indistinguishable species of leaf-hopper communicate and you will see how they select a suitable mate without showing any interest in the seemingly identical creatures living nearby. A bat, screaming ultrasonic pulses to find its way in the dark, may be met by similar jamming vibrations emitted by the very prey it wished to hunt. Bees will build with brick, and wasps make paper.

Some of the senses in the insect world are a refinement of abilities we too possess. Most flying insects have two pairs of wings, but the diptera (two-winged insects) have specially adapted hind-wings. They are no longer capable of contributing to flight, for they are highly modified and look like tiny pins stuck in each side of the fly's thorax. These protruding organs, the halteres, are most easily seen in larger diptera, like *Tipula*, the daddy long legs or crane fly. It has always been said that they are organs of balance. Recent research has shown that the blue-bottle fly *Calliphora* has 335 strain receptors at the base. They pick up the stresses and strains in the exoskeleton of the fly, so it may be that these halteres hold further mysteries for us.

Other senses are bewildering, for we associate them with more specialised creatures. The monarch butterfly of North America flies 3000 km (1850 miles) south each autumn to over-winter in Mexico. They know exactly where they are heading, for they all end up clustering on crowded 'butterfly trees', millions of the insects all arriving at the right place and at the right time. The butterflies flutter like leaves flying in the breeze, blown about by turbulence, flying unsteadily this way and that. Yet they can use pin-point navigation that any human pilot would admire, arriving at a particular spot after a journey equal to the width of the Atlantic.

Insect miracles

The notion of senses in insects takes us into alien territory. With their angular exoskeleton they look like archetypal aliens, and it is hard to imagine them showing any form of intelligence. What we can search for is behaviour which adapts itself to the unexpected or novel situation. Insects and their allies are well able to work out a new strategy to deal with unforeseen circumstances.

Insects, and other arthropods from spiders and mites to crabs and lobsters, share so little with which we can identify, yet many of them form complex and organised social communities which parallel human society to a remarkable extent. A whole ants' nest is like a single organism, with each individual contributing to the good of the whole. We tend to assume that there are separate castes of ant, one for each task in the nest. Not so. There are ten times as many tasks within the colony as there are types of ant to carry them out. Ants can learn new abilities, and modify

their behaviour in the light of the demands of the moment. I
have watched ants climb over each other, forming a bridge, for
the transportation of prey across a gap. To assume that there is
an inbuilt reflex pre-ordained for this very eventuality is incon-
ceivable. The ants adapt their behaviour to fit the circumstances.
For many ants, the need to form a bridge might never arise. In
those where it does occur, no two 'bridges' are ever going to be
the same. These tiny insects are making decisions, working out
what's best.

Canopy ants are an extreme example of this ability to work as
a team. These tropical creatures form a tent-like structure by
holding on to each other. Underneath this protective shield the
colony feeds and reproduces. From time to time the entire
colony moves to a new site where there might be more food.
The ants climb down, move to the new location, then re-erect
their living canopy. What explanations can we offer for this?
There are four possible propositions:

1. The ants are pre-programmed to carry out this task, like
   mechanical robots. That won't work. The amount of brain
   capacity the ant would need to store every possible combi-
   nation would be far larger than the ant itself. In any event,
   no two canopies are ever the same; decision-making and cal-
   culation are an inevitable part of the process.

2. Canopy ants are governed by a force field. I see no reason to
   invoke this clumsy concept, though it certainly fits the facts
   better than the previous proposition.

3. God has ordained how the ants behave. This does not seem
   to be necessary, either, even though it may well satisfy those
   with faith in such matters. This theory, however, still remains
   more convincing to me than proposition 1.

4. The ants use their mental abilities to work out what's best. They conceive of their problem, undertake the requisite problem-solving, and act as a community to carry out what is appropriate. In their own fashion, ants think.

God (and force fields) are governing agencies to which many people subscribe. Proposition 1 – that an inbuilt automatic response covers every possible eventuality – is both fantastical and superfluous. Proposition 4, where I propose that the ants use their brains, not only fits the facts but eliminates any need to seek the supernatural. The pre-programmed principle of Proposition 1 is an invention accepted only because it allows humans to retain their unique superiority. Forget it; it's an illusion. The human brain is certainly gifted beyond comparison, but not all of its abilities are confined to our single species.

The caddis fly

Insects have a wide range of finely developed senses, and can undertake surprising tasks. The caddis-fly larva constructs a house for itself out of inert materials it seeks out on the sandy bottom of the water in which it lives. Many of them can be recognised by a fondness for one particular kind of constructional element, whether tiny stones or fragments of plant stems. The larva starts with a tiny case of particles held together with silk which it secretes. The first case often has a hinged door at the front, which can be closed when necessary, and is often a single piece of leaf or large grain of sand. As the larva grows, it adds to its home. You can watch them scurry around the bottom of a pond or stream, looking for the particles they prefer. Some of them cut from the leaves of water-plants segments of a carefully controlled size using their specially adapted mouth-parts. Sometimes a larva will take over an empty case, and adapt it for its own use.

Orthodox explanations deny that there is any 'thought' in this. However, some observations suggest that a caddis-fly larva knows perfectly well what it is doing. Some species build cases with a characteristic serrated outline. This results from the way they cut fragments and glue them together, for the outer edges protrude and create the jagged contour. Now, what happens if we smooth the case off artificially, by cutting away the rough edges? This is where the larvae show that they are more than mere automata. Although there is no natural manner in which such trimming could be done, the larva immediately sets out to rectify matters. It returns to its source of supply and cuts fresh fragments, which it glues in to cover those that have been smoothed off by the human experimenter. Within an hour or two, the original contour has been restored. The larva within cannot have been greatly troubled by the trimming; indeed, it could be argued that it improved the look of the case.

However, the larva knew that the case was not meant to look like this. It is supposed to have a jagged contour, as an aid to camouflage, and the topographical contour of the case varies from one species to another. Accordingly, the larva repairing a damaged case makes accurate assessments, finds the raw materials, adjusts its 'instinctive' assembly technique and creates a new profile that restores normality. It is hard to dismiss all that as a series of instinctive reactions, all pre-programmed into the larval mind. Evolution is unlikely to equip such creatures with automatic reactions to every unforeseeable consequence. A form of analysis, examination and thought about the problem fits the realities far better than some extreme mechanistic explanation.

The insect edifice

Other insects create much larger, spectacular edifices and have to process the raw materials to make them. Examples range from the huge termites' nests to the delicate spheres of paper in which solitary wasps build their nests. The termites themselves show that they have a magnetic sense when they plan their communities. Their vast nests are built with an axis related to the direction of the earth's magnetic field. If termites are kept in an iron enclosure (which shields them from the earth's magnetic influence) they lose their sense of direction. In experiments in which termites are kept in the presence of an artificial magnetic field, they construct the nest axis oriented to that of the magnet, rather than according to the points of the compass or the position of the sun. Their large nests, hard as concrete, are made by cementing together grains of sand, each exit being carefully constructed to maintain thermal currents of air which ventilate the interior. Apart from their magnetic sense, termites have considerable abilities as architects. They even use insect repellant to control infestations of their nests. For over a century we have used naphthalene to make mothballs, confident that it will prevent clothes moths from laying their eggs in woollen garments. Termites do the same. Researchers have sometimes noted a faint odour of mothballs in termites' nests, and they have now been shown to use over 200 micrograms of naphthalene per kilogram of nest material to keep the colony free of unwanted insects.

The industry undertaken by some leaf-cutter ants is close to farming. They excavate large underground nests which the colony inhabits. Workers go out foraging for leaves which they cut with their jaws and bring back to the nest. These leaves are used to grow colonies of fungi, enzymes from which can digest the cellulose cell walls of the leaves and render them suitable for eating by the colony. The energy of the fermentation also sustains the ants, keeping them warm. The worker ants masticate the fragments of leaves to mix them with the fungus spores, which has the effect of optimising the fermentation in the underground fungus culture. When winged females leave the colony to found new ants' nests, each collects a small portion of the fungus and transports it in a pouch in her mouth. Once mating is over, the fertilised female ant uses this fragment to start a new fungus garden. The garden is vital for the ants' survival; without care and the continuous farming and feeding of the fungal colonies, the ant colony is doomed. These ants are indulging in an agricultural enterprise which they systematically maintain.

It now seems that the fungi may actually be in communication with the ants. In an experiment at the University of Southampton, a captive community of leaf-cutting ants was offered vegetation which had been treated with a systemic fungicide. The ants collected the food as normal, adding it to the fungus garden. Within two days (before any visible change had occurred to the garden) the ants had stopped collecting the contaminated vegetation. When uncontaminated vegetation of the same kind was provided, the ants all ignored it. The research workers concluded that the fungus had a means of communicating the fact that it was under threat. It is equally clear that the ants seem to have been able to communicate a warning about the danger of this kind of food to the other foraging ants in the colony. The colony is the collective home to all the ants, of course, and we should not be surprised to learn that they are very efficient at detecting problems and communicating the danger to their fellows.

Weaver ants provide perhaps the ultimate example of co-operative construction, in which teams of workers combine their efforts in the interests of a grand constructional project. These insects build their nests by sticking together leaves with silk. A single ant cannot reach to join the leaves, nor do the ants themselves produce silk. Instead, they work as a team. Rows of insects seize the edges of leaves to bring them together. The distances are usually too great for ants to bridge on their own, so they climb over one another, each holding onto the ant in front, until eventually the gap is bridged. By careful manoeuvring, the massed ranks of ants manage to move the edges of the leaves closer, until they can come together.

Adult weaver ants do not produce silk. Instead, they bring silk-secreting larvae from the brood-chamber and hold them gently in their jaws. They signal to the larvae when it's time to go, and then apply them like a tube of glue, first to one side of the join and then the other. In this way the gap is secured with threads of silk like delicate embroidery. The employment by adult ants of their larvae is close to the use of tools by these insects, and the cooperative labour in bringing the distant edges of leaves together shows much planning and on-site decision-making. To dismiss this as mere programmed action is immensely short-sighted, for nature reveals much thought and careful coordination at many different levels of life.

Insects and cohabitation

Ants have developed a highly evolved social system. They communicate through many channels – through scent, chemical messengers, vision and possibly through sound. Ants can use their own ant language well enough to carry out the complex affairs of life in a community. Could other organisms learn to speak like an ant? Could we? I have no doubt we will learn to understand their communications. Ants are not the only organisms to speak their language, for some other species have already cracked their code.

These are the many creatures that live inside ants' nests. Were a 'normal' invader to try to settle in an ant colony, it would be quickly identified and attacked. Not these creatures; they are tolerated by the ants. Indeed, they are nurtured and groomed by them too. Thousands of different species live with ants, including beetles, wasps, flies and mites. One example is the beetle *Atemeles pubicollis*, which is found across Europe. It co-exists with the wood ant *Formica polyctena*, which raises the beetle larvae in the ants' nest. Each beetle larva uses chemical attractants to fool the ants into cooperating with it. Radioactive labelling has been used to trace the movement of marker sub-stances from the larva to the ant. If larvae are presented to the ants, they will be taken into the nest without hesitation. However, if the larvae are coated with a thin film which prevents any scent from escaping, the ants reject them. If a small portion of the skin is left uncovered by the insulating layer, the messen-ger molecules which diffuse out are picked up by the ants, which willingly transport the larva into the nest. They will also react strongly to dummy larvae which are treated with the exudate of the real thing. Clearly, these ants need to scent the larva to recognise it as belonging to their community.

The young ant larvae are fed by regurgitated food from the foraging adults which return to the brood-chamber after a spell out in the forest. The adult leans over the brood-chamber to inspect the larva. As soon as the larva detects the presence of a grown ant, it draws itself upwards and signals to the ant that it wishes for food. It does so by tapping on the mouth-parts. In response to this signal, the adult passes some regurgitated food to the larva. Not surprisingly, the beetle larvae do the same. As the adult ant leans over, the beetle larva reaches up and taps the ant with its own mouth-parts. The signal is clear, and the ant regurgitates some food for the beetle larva. Using radioactive tracers it has been shown that the amount of food consumed by the beetle larvae is disproportionately large, and as they mature they add nearby ant larvae to their diet. As a result, brood-chambers containing ant larvae have normal populations, whereas the chambers in which the beetle larva thrives will con-tain no ant larvae at all – they will, however, be surrounded by attentive ants busily providing food.

As the season goes on, the colony begins to wind down for winter. The *Atemeles* larvae have pupated and hatched, but the young beetles are not yet ready for adult life, for they are still immature. Since the wood ant colony is not supporting them properly as temperatures slowly fall in the autumn, the young beetles set off to find a more accommodating host. This is *Myrmica*, an ant which maintains its brood colonies and its foraging activities throughout the winter. In this nest the beetle will be perfectly at home, and so it makes plans to move out.

Film taken within the colonies reveals how the young *Atemeles* approaches one ant after another, drumming on the ant to indicate urgency, and then meaningfully touching the host's mouth-parts to plead for food. With enough urgent attention the ant will regurgitate a supply of food for the beetle. It moves on, repeating its signalling, until it is replete with food. This extra supply gives it enough reserve for the transfer to a new colony of very different ants.

When it is full of food, the beetle quits the now-quiet wood ant nest and sets out to find a colony of *Myrmica* ants. These do not live in woodland, but in open fields nearby. The beetle moves to the edge of the wood, guided by its sense of sight. During the two weeks after it leaves the *Formica* nest, the young *Atemeles* can scent *Formica* colonies on the air. Before this time, and afterwards, the sense of scent is absent, and it has a purely functional role in short periods of the beetle's life. The clearest indication of the need for it to scent the ants on the air lies in the observation that the beetle remains perfectly still as long as the air is not in movement. It needs scent, carried on the breeze, to locate its new host.

Once it has found a suitable nest, it moves in for the winter. The beetle approaches the new colony. When it meets an ant of the new colony it taps it gently with its antennae. Then it stands still and raises its abdomen for inspection. The ant tastes

secretions at the tip of the beetle's abdomen, as though checking for identity. That done, it then senses secretions from
glands situated in lines down either side of the beetle. When all
is done, the beetle lowers its body and it is carried shoulder-
high into the brood-chamber where it immediately takes up
residence.

The adaptation means that *Atemeles* must be trilingual: as
well as its own codes of communication, it understands the differing languages of two very different ants. There are other
beetles which inhabit ants' nests, and some of those also change
hosts with the seasons. Some species have a less intimate association, for they graze on the discarded wastes of an ant colony,
and pick up food the ants no longer want. Although they are
not attacked by the colony and ejected from the site, they are
killed if they ever find themselves inside the nest. Other beetles
act as wayside opportunists. These highwayman beetles find the
regular route used by an ant colony and linger as the ants pass
by. Approaching a foraging ant on the homeward path, they will
tap it in the approved fashion and trick the ant into releasing
some food, which the beetle quickly seizes. In most cases the
ant realises it has been tricked, and attacks the beetle. There is
no time to flee, so the insect simply clamps itself down on the
ground and relies on its specialised carapace to keep the attacker
at bay. The attitude adopted by the beetle as it requests suste-
nance, body lowered to the ground, head raised upwards as
though begging for food, is compliant even to the human
observer.

Some of the lodging beetles are adapted so that they look
similar to the ants among which they live. This is not to make
them inconspicuous within the ant colony, but to prevent their
being singled out by predators from outside. This can be simply
proved, for if something is done to alter the shape of mimicking
beetles, or those (like *Atemeles*) which are quite unlike their

hosts, the ants take no notice of the change. They rely not on the sight of the beetle, but on its scent and its tactile behaviour. There is clearly a language here waiting to be understood. Beetles can do it, and so should we.

Evolutionists point out the complexity of the developmental stages such a process has to undergo:

○ The would-be parasite has to develop senses with which to recognise the ant's communications.

○ It has to learn the signals and understand the language of communication.

○ It must then have evolved the exact secretions which allow it to respond to the ant.

○ Finally, it needs to have developed an ability to mimic the tactile and tapping signals the ants use within the nest.

In short, these beetle larvae need to have learned the language and to have copied it perfectly. The conventional view is that this evolves through survival of the fittest, those that could not get close to the ant's way of living being eliminated by natural selection. This would necessitate a belief in the beetle larvae evolving a random array of tactile signals, just one of which happened to match the ants' expectations. This is a notion of ludicrous complexity, if we have something simpler that fits the facts. It is clear to me that, just as other animals can learn from experience, the parasitic beetles have simply learned what to do. Inborn patterns of behaviour exist in every species. The challenge for the new millennium is to find how a learned pattern of behaviour can be incorporated into the hereditable basis on which future generations will be predicated.

Insects and intelligence

Wasps also navigate, and regularly fly considerable distances to find suitable wood which they then collect and macerate to produce paper pulp. They extrude delicate, carefully constructed shapes and the result is a wonderful nest roughly the size of a football. The process is lengthy and complicated, yet each insect manages not only to process the wood it gathers, but to fit its contribution into the whole.

Other flying insects make nests which, if done by humans, you would describe as the product of high intelligence. Sand wasps excavate a burrow with a chamber at the far end. The sand wasp then flies off to find a caterpillar that will just fit into the burrow, paralyses it and hides it in the chamber where she lays her eggs. The caterpillar survives, and is eaten slowly by the growing sand-wasp larva. As the sand is removed from the burrow, the wasp makes sure it is widely scattered, so there is no hint to the outside world of the excavation she has done. To close the entrance she uses a stone of carefully chosen dimensions, and will fly a considerable distance to find the exact one she needs. Meanwhile, she watches where she is flying and returns unerringly to her nest, even though it is effectively camouflaged from other intruders. Her sense of sight, memory and ability to navigate compare favourably with anything human senses can achieve.

Mason bees carry out extraordinary feats. They construct brick enclosures by cementing together particles of sand into small pellets, then gluing them to the surface of solid rock to form nests. The finished construction is then coated with rock dust and sand to disguise it. It ends up as exceedingly hard, and the bee that eventually hatches within has to excavate its way out. Other bees lay their eggs in empty snail shells, filling the open end with food reserves topped with leaves and tiny pebbles, which leave the occupant safe and secure but able to breathe through air spaces carefully left in the structure. The shell is then disguised with blades of grass or pine needles so that it will not attract attention. Other bees use these snail shells for their own young, and commandeer the partly completed shell like a cuckoo in a bird's nest.

Insect language

Do insects communicate? Some simple observations prove that they do. One day you might see an ant in your kitchen. Let us suppose that it has found its way to a sticky jar of honey. Next day ants will be there in force. They follow the same pathway, having established a line of communication from the nest to the new food source. If the line is broken they will try to find another way in. Remove the jar, clean up the honey and, if there is no more food, the train of ants will dwindle and disappear.

That is a straightforward example. The ants must have been following a scent trail, for which the most basic chemical communication would be necessary. It is not so simple for insects that fly, and do not have any contact with a trail. This is the case with the honey-bee, *Apis*. A demonstration proves the point. Set a little dish of sugar-water on a ledge, or in the garden, and watch it for a while. In time, if you are not too far from a hive, a honey-bee will find it. It settles down at the edge of the liquid and starts to drink it. You will probably see the bee spend a while grooming itself, for it may have sacs of pollen and stores of nectar ready to take back to its hive. When all is done, it flies off into the distance. Your initial observations are over at this point; now all you have to do is wait for half an hour or so. By then, there is every likelihood that the dish will be swarming with bees. Within an hour there may be hundreds of them trying to get at the food. The first bee to find the supply has returned to the nest and explained exactly where it is to be found – distance, direction, even how rich a food source it is. The bees which have invaded your little dish of sweet food were told where it could be found, and were told by the exploring bee which first located the prize.

In similar scientific experiments, scientists have moved a dish progressively farther away from the hive, and watched the way the dance changes as foragers convey their findings to the eager onlookers. It has been noticed that, for the first move or two, a considerable period of time goes by before the first bee finds the dish and sets off to warn the others. From then on, the dish may be moved far larger distances, and the bees find it with little delay. The bees understand that the source of food moves away in a straight line once in a while, and (if it has disappeared) they immediately work out where to look. Nothing like this happens in nature. We are witnessing the bees learning about a new food source and quickly acquiring a technique to deal with it.

Bees communicate in many ways; they recognise each other, and their favourite food sources. These communal insects manage to communicate when problems arise, when a predator threatens a colony, or when the young need attention. In some cases they speak a language which humans can follow. The best-documented example is the language of the honey-bees, for that was painstakingly unravelled by the Austrian scientist Karl von Frisch (1886–1982). His observations began with an inquiry into how bees communicate their discoveries. He made painstaking observations of the way bees behave in glass-sided experimental hives, which revealed the astonishing dance of the honey-bees. The explorer bee returns to the hive with knowledge of the whereabouts of a new food supply. She starts her dance by running across the surface of the honeycomb, waggling her tail as she goes. At the end of this run she stops the waggling and returns in a flattened semicircle to her starting-point, from where she repeats the performance. This time she returns in a loop in the opposite direction to the first one, and repeats the waggling run once more. She keeps dancing like this as onlookers gather and study what she is doing.

Von Frisch showed that the angle of the direction of the waggling run is related to the compass bearing of the source of food, while the intensity of the waggling corresponds to the amount of food to be found. Distance is a function of the frequency of the waggling movements. He found that, if the food is 300 m (1000 ft) distant, the bee waggles her tail 15 times in the 30-second dance. If the food supply is twice as far away (600 metres or 2000 feet), the number of waggles drops to 11. When food is close to the hive the bee changes its performance, and the circular dance it performs instead signifies a food source within a few minutes' flying time.

Interpreting the waggling dance was relatively easy, once its significance had been recognised. In warm weather, when it is too hot to dance inside, the foraging bees perform in the open air, often on a flat surface like the landing-strip in front of the hive. Under these circumstances, the bee's waggling path points exactly in the direction of the food source. Inside the hive, the bee uses her sense of gravity to establish an exact vertical. She then uses this as a reference point for the direction of the sun, and the angle of the waggle track to the vertical corresponds to the angle of the sun to the path which takes the bee to the source of food. Thus, if the angle to the food is 35° to the left of the sun, the bee's waggle dance will be 35° to the left of the vertical on the hive. Other bees follow the direction and amount of agitation in the bee's dance. They memorise what they learn, and set off to find the food. If little bowls of sugar-water are set out closer to the hive, or farther from it, the new arrivals ignore them and fly straight to the supply taught to them by the explorer bee. This is clearly communication and is a form of expression which humans have been able to interpret.

Do all bees speak the same language? The waggle dance is a universal language for honey-bees, but it has different forms in different strains. Research has shown the differences from one type to another, and von Frisch aptly called them 'accents'. One example is the way the round dance is used instead of the waggle dance. The Indian honey-bee *Apis indica* uses the round dance only for food closer to the hive than 3 m (10 ft). The Italian honey-bee, *A. mellifera* var. ligustica, resorts to the round dance for distances up to 15 m (50 ft), while the Austrian honey-bee, *A. mellifera* var. carnica, uses it for distances up to 85 m (275 ft). There are differences in the nature of the waggle dance between these species, for they use different numbers of waggles to indicate distance. Are the languages innate? There can be little question about this aspect, for bees reared in isolation respond immediately to a colony when they are introduced to the hive for the first time. They understand the language of their own kind and have no hesitation in responding correctly. But wait – what of the Italian and Austrian bees? They are varieties of exactly the same species, *Apis mellifera*, and can be inter-mixed and even cross-bred without difficulty. What would happen to the behaviour of the hybrids? Von Frisch decided to try it out.

The Italian honey-bee has a dance of its own, the sickle dance, which von Frisch also observed. It is a strong and pronounced 'accent' within the language of the honey-bees. It sends the wrong message for the Austrian variety of the bee. Since these two varieties are of the same species they can be mixed in a colony without many problems, and it was seen that if Austrian bees watched an Italian bee's dance, they set off in the right direction but the sense of distance was lost. An Austrian bee taught by the dance of its Italian cousin flies too far to find the food source. Observations on hybrids proved much as one might predict. The hybrid bees manifest the characteristics of either parent. Italian bees have yellow markings, and these hybrids were seen to do the characteristic sickle dance on almost every occasion. Those more closely resembling the Austrian parent's body colour danced the Austrian version 96 per cent of the time. On the other occasions these Austrian-type hybrids tried a dance rather like the Italian sickle dance, but lost the orientation to the sun in the process.

There are other species of honey-bee, and they all have their own version of the dance. Most primitive is the dwarf bee, *A. florea*, so small that it looks at first like a flying ant. It builds a single comb, smaller than a saucer, which hangs down from an upper branch of a tree. The bees returning from foraging expeditions always alight on the top of the comb, where many of the bees in the colony are to be found waiting. They perform their dance in the light, using the round dance for food up to 5 m (16 ft) away and a version of the waggle dance for food that is farther away. The rate of waggling is proportional to the distance, though it is slower than for the common honey-bee.

So much is there in common between the European honey-bee and the Indian bee that for many years they were believed to be varieties of the same species. We now classify the Indian bee as a distinct species, *A. indica*. Like its European cousin, it is widely farmed for its honey. These bees use the round dance for distances up to 3 m (10 ft) and from then on describe their finds using their version of the waggle dance. Whereas the dwarf bee uses the line of sight for navigation, the Indian bee (like the European species) builds its combs in the dark and therefore uses the vertical to indicate the direction of the sun. Its rate of waggling is similar to that of the dwarf bee, and is much slower than the European species. *Apis dorsata* is the giant honey-bee. Like the Indian bee, it changes from round to waggle dance at about 5 m (16 ft), but it waggles at about the same frequency as the Italian bee.

The bees thus group together, with the Indian, the giant and the dwarf honey-bee all restricting the use of the round dance to food sites close to the hive. It is believed that the dwarf honey-bee is the most primitive type, in developmental terms, and the dance tends to substantiate this view. This species cannot transpose the horizontal-pointing dance on a flat surface to a vertical dance. If the comb is up-ended, they try to find a flat place on which to perform. If they are held in captivity in a chamber with vertical sides they are unable to dance at all, but try to find somewhere flat.

Transposing the horizontal dance to one on a vertical surface, where the downward pull of gravity is used as a substitute for the line of sight to the sun, takes a greater degree of specialisation. For the dwarf bee, this step is too much to take. The transposition is instinctive in many insects. If a dung beetle is observed in captivity it may set off in a straight line that is at an angle to the light in the room. Now turn off the light, and up-end the flat surface on which the beetle is walking. It is now at 90° to the horizontal plane. The beetle continues to crawl in a straight line, and the angle of its path to the vertical is exactly the same as the angle of its previous path to the direction of the light. Many other insects can make this transposition from modelling through light to navigating by gravity.

One important matter must be addressed: how do bees observe the dance? When making films of the phenomena, scientists spend much time lighting the glass-sided hive in which the bees live. But a normal bee-hive is inky black inside. We now know that the observing bees that cluster round a dancing forager are detecting the movements of her abdomen by the pulses of air pressure set up by the movement. While they use their eyesight to observe the outside world, a new sense – detecting air pressure – is developed to tell them what goes on when bees dance.

The use of the sun is of crucial importance. Bees have compound eyes which are so geometrical and precise that they must surely help in maintaining a line of sight to a source of illumination. Bees maintain an accurate biological clock, for the direction in which they fly is not modified as the sun changes position. They are constantly resetting their direction-finding abilities as the sun moves across the sky. Many flying insects live in desert conditions where a sense of sight would be of limited value. The shifting sands mean that the landmarks around a nesting site are continually changing, and can look very different during the return journey from how they looked when the insect set off. For communities of bees, the sense of navigation is seen at its most important when an overcrowded bee colony sends out queen bees to establish new nests. Workers fly off to find cavities in which the new nest might be built. They inspect them all carefully, climbing into the far crevices, before they all return home. At this point they perform dances and communication sessions, clearly deciding which shall be the new base. Once a decision is made, the entire colony with their queen flies off to the chosen site and there begins to establish the new colony. All the bees know where they are going, and what they have to do when they arrive. Much complicated communication takes place between the individual bees if foragers are to reach a decision and convey it to the rest of the swarm. Even in the primitive and alien world of insects there are complex communications going on, and decision-making is plainly apparent.

The insect brain

In all these examples, the insects are having to make extraordinary leaps in mental ability. They are predicting their actions, modifying their behaviour, communicating their conclusions and (in the case of the dancing bees) translating mathematical certainties into a coded form of dance, which onlookers will understand. Conventional explanations are that these insects are simply pre-programmed, but their brains, containing several thousand neurons, could not embody sufficient exigencies to meet all possible occurrences in their daily lives. In their own way, these insects are solving problems and thinking. They are communicating in an insect's way, not in the ways of humans; but they are communicating deliberately and in detail.

As we have seen, the vibrations created by insects are used by them to provide information about the surrounding world. Whirligig beetles of the family Gyrinidae spin around crazily on the surface of a pond. They move at high speed, taking random curling pathways across the water. Careful observation shows that they always maintain a set distance from one another and never collide. The apparently wild gyrations are far more controlled than they appear. Each beetle keeps its antennae in contact with the uppermost layer of molecules of the water on which they spin. Vibrations set up by all the insects are detected by the beetle, which creates a picture of the community and the trajectories of each one. These beetles have the ability to sense where the other beetles are, and where they are heading. The capacity to model the movement, and to know where it is taking each individual, shows a high degree of mental ability.

Spiders similarly rely on vibrations to tell them what goes on in their neighbourhood. The orb spiders which spin the webs we find in gardens and fields use the webs as an extension of their bodies. The web is secreted by the spinnerets at the rear end of the body, and the spider uses its legs to measure the web and the distance between each silken thread. Once the web is finished, the spider retains contact with it by resting its forelegs on a signal strand running from the centre to the spider's lair. A fly trapped by the web sets up vibrations which are characteristic of the species. The spider responds accordingly. Spiders have good vision with their eight eyes, scattered across the head, and can recognise their prey once they are close to it. There is the view that spiders simply act like little robots in response to pre-determined stimuli, but they belie this by showing a clear ability to use their brains to adjust their behaviour in the light of the unexpected. If a fly is ensnared in the web, the spider darts out along the threads towards its prey, inspects it, then immobilises it with a nerve toxin secreted by glands near the biting mouth-parts. Once the fly is still, the spider secretes silk and turns the fly in its forelegs, wrapping it up like a mummy, ready for future consumption. Suppose an experimenter holds a living fly near the spider by means of forceps. The spider inspects the fly, then seizes it and poisons it.

What if a fresh (but dead) fly is placed on the web and agitated manually to attract the spider? The spider runs up to inspect the fly, and sets about wrapping it in silk. The spider does not pause to waste time biting a fly which doesn't need it, even though the arrival of a freshly dead fly in its web is something it will not experience under normal circumstances. There is evidence here of a mental ability to work out what's best, even in an entirely novel situation.

The rates of vibration detected by insects and their allies in these situations are low, but some insects can detect exceedingly high frequencies. Many species can hear vibrations that lie outside human hearing. We have seen that, when we were young, our ears could probably detect high, hissing notes up to 20 kHz. As we age, this upper limit comes down to 10 kHz, or even lower, though this makes little real difference to our perception of everyday sounds. In some cases of deafness, a reduction in the upper limit to a few hundred Hz is as significant as actual hearing loss. The result is to make sounds muzzled and indistinct. Sibilant consonants are a key feature of speech, and such sounds are lost in this condition. 'Tea' and 'sea' may be hard to distinguish, for example, and meaning is easily obscured. These findings remind us of the crucial importance of vibration detection in our own lives. Insects inhabit a world far richer in sounds. While we find it hard to hear anything above 20 kHz, and hear sounds above 12 kHz as little more than a high-pitched hiss, insects can hear 50 kHz without difficulty. Nocturnal moths, which are hunted by bats using echolocation, set the record for they have been shown to respond to frequencies higher than 200 kHz. This gives them a range of hearing covering a sound spectrum some ten times broader than our own. They rely on this sense to avoid predatory bats in their vicinity.

Vibration sense is also used by insects to communicate, as well as to perceive. Crickets, cicadas and grasshoppers produce astonishing sounds by stridulation, in which they rub parts of their bodies together and set up mechanical vibrations. Crickets can agitate their specially adapted wings 50 times a second, and as they rub together they produce a whirring sound. The fastest they can rub their wings back and forth is 50 or 60 cycles a second, but this produces a low humming noise and sounds of this pitch do not travel far. To overcome this physical limitation, crickets have evolved grooves across their wings. Each wing runs a sharp plectrum-like protrusion across the array of grooves, like a finger-nail along a comb, so that a far higher frequency is produced. In this way a signal of perhaps 4000 Hz is generated from the movement at 50 Hz. A triangular portion of the wing, the aptly named harp, resonates in sympathy with the frequency, and the beam of sound is projected for considerable distances. Some species, the bush crickets, have refined this to such an extent that they produce a signal in the ultrasonic range. They can hear each other remarkably well, whereas mere humans are unable to hear anything other than the 50 Hz of the rubbing of their wings.

Producing the sound is one thing; detecting and analysing it is a very different task. The sound is received by the paired ears, each carried on the insect's legs, and relayed to two giant cells – the omega cells – inside the body within the first thoracic ganglion. They process the pitch of the sound and the relative intensity with which it is detected by the ears on the two forelegs. The cricket's brain is able to process this result, and the insects can detect the direction of an incoming signal to within 5° on the compass. Cicadas keep their paired sound-producing organs, the tymbals, in continuous vibration by muscles which cause them to resonate. Some cicadas make sounds so monotonous and so loud that they sound like poorly lubricated machinery churning continuously.

The Chinese, who keep crickets in captivity, find the sound of these creatures so appealing that they have bred them for the beauty of their song. Sound is one of the most important cues for these creatures, certainly more important than sight. The simple experiment which proved the point set a male cricket under a glass container in full view of females ready to mate. The females showed no interest. If the sound of the male was picked up by a microphone and relayed through a nearby loudspeaker, all the virgin females quickly congregated around it. They find sound more instructional than vision.

The ears of insects are varied, and are not always on the head. In grasshoppers, for example, they are found at the knee-joints on the forelegs. As the limbs are raised and lowered they give the grasshopper a full, three-dimensional picture of sound sources in the vicinity. Male mosquitoes detect the wing-beat of females through their feathery antennae, which resonate at the frequency of the female's hum. If a tiny weight is added to the male antenna, the mosquito ignores nearby females, for its tuning frequency is disturbed and its feelers can no longer vibrate in sympathy with theirs. Until the male mosquitoes are sexually mature, the fine hairs along their antennae lie against the shaft, and so do not respond to the sound of the female wing-beats. Once the males mature, the hairs become erect and the insects can respond to the females. Similarly, until the females are sexually mature they produce a hum of too low a frequency to be in the range detectable by the males. As they become fertile, their wing-beats speed up to produce the high-pitched hum to which the male antennae can respond.

Insects conduct courtship rituals. Fruit-flies meet on the ground, and spend much time getting to know each other. The male taps the abdomen of the female with his forelegs. He struts around her, exhibiting his body markings, flicking his wings, vibrating his body. She watches all this intently, turning her head and paying close attention. Should she choose to reject him, she vibrates her wings as though ready to fly off, but holds firm with her feet so that she stands her ground. At this sound, the male cocks his head and flies away. There is an elaborate series of courtship conventions in all this display. The flies recognise each other as individuals and clearly use sight, sound and scent to explore a possible relationship. It is the male which makes the presentation, and it is up to the female to make the final decision. Parallels elsewhere in the animal kingdom are not hard to find.

Crude mosaics?

The compound eye of insects has always been regarded as if it acted like beaded window glass, providing a crude impression of the world. The eye of a fly was one of the first objects ever examined under a microscope, and the first experiments to obtain an image through a fly's eye were published in 1665. From that day to this we have imagined that a fly obtains a crude and unfocused image of the world, which is a groundless and absurd conclusion. Each compound eye contains large numbers of separate optical units, the ommatidia, arranged in a rounded cushion. There are up to 30,000 ommatidia in each of a dragonfly's eyes, for example. A single ommatidium has much in common with a conventional mammalian eye: there is a cornea, which focuses the primary image, a lens behind it, and cells forming the retinula which gather the light. Conventional theories hold that the fly sees an image made up of dots of light corresponding to each ommatidium, making a picture like a crude half-tone plate from an old-fashioned newspaper. Experiments suggest that this model cannot explain the acuity of an insect's vision: its sense of sight is massively better than that.

Tiny implants can pick up the electrical signals generated when an ommatidium is stimulated. The mosaic theory assumed that a fly's eye would accept light over a field about 2° to 3° across. Analysis of the nerve signals disproves that idea, for each ommatidium could respond to light across a far wider field, up to 30° across. The classical theory implied that an insect eye could distinguish separate points that subtended more than 2° to the eye, but the measurements prove that insects could actually distinguish points no more than ½° apart. The idea that a compound eye could produce only a simple mosaic image is a human attempt to downgrade an insect's abilities, and to retain the sense of effortless superiority. It is clear that each ommatidium provides an excellent image. I have no doubt that the brain of the insect integrates the separate images into a single detailed view of the surrounding territory. As anyone who has tried to catch an insect (or swat a fly by hand) can testify, insects have excellent vision and can see details very well indeed. Most insects are between 1 and 10 mm (about ½₀ to ½ in) in size, and they would need good vision to see each other, apart from anything else. We need to accept the simpler fact – insects have excellent eyesight.

The insect eye perceives colour, and has good vision into the ultraviolet, which is invisible to humans. Many flowers (and a few insects) have markings that are ultraviolet, and patterns which attract insects are often undetected by the human observer. There are ultraviolet nectar guides on many flowers, which help a foraging insect to home in on its target. Insects see best at the blue end of the spectrum, and may see a brightly coloured landscape by distinguishing tints which we would find indistinguishable. Sight is only one of the senses insects use to navigate around the world. Many of them have delicate organs of balance, and their hearing allows them to pick up auditory signals from one another.

There are some species which human zoologists cannot distinguish on sight, but which can tell each other apart by the calls they emit. One current subject of interest is the Indonesian frog-hopper, which has important economic consequences in the rice fields. By analysing the spectra, taxonomists can tell the species apart and learn about their social lives. Thus, by interpreting the secret senses of insects, we may yet find a way to prevent insect pests from damaging the fragile enterprises on which human societies depend.

Molluscan intelligence

The most intelligent of all molluscs is the octopus. It is a responsive creature, and adapts well to live in captivity. In an aquarium it will adapt to living within a compartment with bricks at one end, emerging to collect food offered at the other end of the tank. If a crab is presented by the keeper, the octopus comes out, seizes the crab, then takes it back to its lair to incapacitate and devour it. An octopus can be trained, like many more familiar animals. In early experiments a pulse of electricity was administered to an octopus confronted with a particular object on the end of a stick. The shock was enough to make the octopus recoil. In a very short while, the octopus would refuse to feed if the object was visible. If the shape was brought towards it, the octopus would shrink back. Only if it was nowhere to be seen would the animal come out to feed. These inhumane experiments were the first to make experimenters appreciate that an octopus could learn. The tests also revealed that an octopus has a good memory, because the aversion would last for months. Much has since been said of the sophisticated brain of the octopus, but it is important to realise that there are more neurons in the arms of the octopus than in the whole of its central brain. Much of its mental ability takes place in the arms, without the involvement of the brain.

The senses of an octopus allow it to make fine distinctions. For instance, it can easily discriminate between a living bivalve mollusc (like a cockle or mussel) and a dead shell that is filled with resin. It can distinguish between two shelled creatures of the same size, one with a ribbed shell and one whose shell is smooth. Given artificial objects to inspect, it can distinguish between grooved and smooth cylinders (though not which way the grooves run). Taste is highly developed in the octopus. It is far finer than ours. Substances like quinine or hydrochloric acid (bitter-tasting), or saccharine and sugar (sweet), can be detected by an octopus if they are more than 100 times more dilute than humans can taste. They also have good eyesight. The octopus eye is constructed in a very different way from our own, but its essential features are very similar to the mammalian eye. There are some curious anomalies in the senses of the octopus. For instance, as well as being unable to distinguish the direction of grooves on cylindrical objects, it cannot distinguish between heavy and light objects.

The ancients believed that the octopus was an ingenious

animal. In the year AD 77, Pliny the Elder wrote that an octopus could insert a stone to prevent the shells of its intended prey from closing. Modern investigations have not confirmed this ancient story.

A typical specimen of the large species from the Pacific, *Octopus dofleini*, weighs over 20 kg (50 lb) and has an arm span of about 2.5 m (8 ft). The record octopus was ten times as big, weighing over 270 kg (600 lb) and with an arm span of 9.5 m (31 ft). The ancient reputation of these creatures for malevolent intelligence has given rise to many stories of the 'kraken' and its relatives. Novelists who have been inspired to tell tales about these creatures range from Victor Hugo and Jules Verne to Ian Fleming and Peter Benchley. For all their familiarity, there remain some surprising mysteries. The largest of all these molluscs is the giant squid, *Architeuthis dux*, of which some specimens exceed a tonne in weight. The heaviest on record weighed two tonnes and had tentacles measuring 10.5 m (35 ft) radiating from a body over 6 m (20 ft) long. Curiously, nobody has ever seen a living specimen. This, the largest invertebrate in the world, has yet to be observed by biologists.

The great group of squid, cuttlefish and octopus features some of the most remarkable means of communication in the animal world. They send visual messages to one another via their skin. The cuttlefish, for example, signals with rippling changes of colour running along and around the body. Some of these are clearly meant as threats, others as courtship signals. Some seem to mimic movement which attracts prey, while others are a decoy to throw a predator off-guard. Within the skin are coloured cells, chromatophores, which can rapidly expand and contract. They are typically of three kinds, coloured orange, red and dark brown. Beneath this layer lie a number of cells which – though they have no colour of their own – reflect light with differing intensities. These throw light back to the observer in a range of colours from silver and gold to green and blue. The mechanisms by which the skin changes its colour are immensely complex, and the coordination of the many cells that participate is poorly understood by science. The principle of changing colour is widespread elsewhere in the animal world. There are many vertebrates which can change colour to suit their surroundings. Flatfish merge into the background by adopting a colour that camouflages their appearance, and the chameleon reminds us that the phenomenon is even found in creatures that live on dry land. By any standards, this is an extraordinary use of senses in the animal world.

The homing frog

Every year, the same frogs find their way to the same ponds. How do they do it? They see, they hear, but above all they can smell the water in the soft moist air that lies close to the ground. Take a frog at breeding time, and set it near two artificial ponds. One contains water from a strange pond, while the other contains water from its home pond. It is the home pond-water for which the frog sets out. The frogs know the water they prefer, and can identify the unique chemical mix that floats free on the air.

Frogs do not have a particularly well-developed sense of smell, but newts and salamanders do. In *Taricha*, the red-bellied newt, vivid coloration is a hallmark of courtship. What attracts the newts to pair off is a more captivating cue by far: the sexy secretions of the female. Even small amounts can be sensed by the males. Many species have specially developed structures to help them concentrate the effect. There are grooves in the upper lip, running towards the nose, which serve to concentrate these odours and feed them to the most sensitive organs of smell. In some newts there are tiny hair-like projections to these scent organs which they dip into the water, sampling for female odours. The nodding movements of the head are known as 'tapping' by those who study salamanders, and it is believed to be a way of taking successive samples in order to see whether the male is walking towards a female. The samples will gain in strength if the animal is heading in the right direction. Others have mucous glands on the chin which are used in a slapping ritual with females during courtship. Not only is it a sign of recognition, but the females seem to find their sexual responses heightened by this behaviour. The slapping overtures made by the male act as an aphrodisiac to the female. All newts have a fine sense of smell. Blind newts can unerringly find their way home to a favourite stretch of water, but sighted newts in which the olfactory nerve has been damaged are soon lost. Some investigators have suggested that newts are far better at finding their way home if they are downwind, rather than upwind, of their home base. This seems to support the argument that they have an acutely accurate sense of smell.

When different communities are in close proximity, newts and salamanders prefer the company of individuals from their home group, rather than adjoining neighbours. It is interesting to see what happens in the breeding season, for then the instinct to join their fellows disappears. Finding a suitable partner for sex then matters more than the need to find safety in familiar surroundings. Salamanders on their home territory do more tapping than they do if they are placed on foreign soil. What is more, they show more frequent escape reactions if they are on strange soil, but seem perfectly at home on ground which smells familiar.

They also seem able to use their fine sense of smell to follow prey. In some experiments, fast-moving prey (centipedes) were left in one jar of water, and slow-moving prey (snails) were left in another. Given a choice, the newts always headed for the water which retained the odour of the slow-moving creature.

This substantiates the view that these animals use their sense of smell to recognise their prey.

Amphibians and their territories

There are many senses amphibians use in their daily lives, and many serve to help them recognise their territory and protect it from outsiders. Salamanders and newts do not have an eardrum, or a middle ear cavity, but they do have a well-developed inner ear and they may use that to detect underwater sounds. They are mostly silent, which makes sense if they are essentially deaf, but newts do occasionally squeak. A recording of squeaking newts which I broadcast in my twenties caused much comment at the time, since most naturalists seemed confident that newts were dumb as well as deaf.

Frogs are different. They have eardrums, and they use their voices for several purposes. They have calls to mark out breeding behaviour and for courtship, and they sit and call to ensure their territory is sacrosanct. Sometimes the call is the best way to tell similar frogs one from another. Two species of frog, *Hyla versicolor* and *H. chrysoscelis*, are indistinguishable to the human eye but have calls that immediately set them apart. The sense of hearing in frogs may be sensitive, but it is also selective: the frequencies they hear best are those emitted by their own species during mating. This helps to restrict the babble of competing voices in the springtime, and many species of frogs have ears that are tuned to the voices of their mates.

Frogs have dialects which vary from place to place. An analysis of their hearing abilities shows that the ears of the various frogs are attuned to hear the different frequencies of each dialect. The frequencies are important. Low frequencies travel farther than higher ones. In this way the lower frequencies of the male's call may be responsible for attracting females from a distance, and then (as they get nearer) they can select the higher frequencies which mark out an individual male. It is on this basis that the evaluation of potential sexual partners takes place. The two genders of frogs are separately adapted for breeding, and males can be identified by a horny pad of thickened skin which is used to grasp a female during copulation. The mating urge in frogs is powerful and over-riding. Will Cuppy, in a book I read as a teenager, said that: 'A decapitated frog starts looking for a mate, in or out of the season. There must be a lesson in there, somewhere.' In earlier days, when amateur experimenters showed a cavalier attitude to the creatures that surround us, it was demonstrated that a male frog retains its hold around the neck of the female even if its body is severed at the level below the forelegs.

Bullfrogs are highly territorial, and use many different senses to detect what's what. They emit loud calls, of course, but also take up threatening postures which deter other males from coming too close. In experiments in which high-quality recordings of frog calls are played back to them, aggressive behaviour is easily evinced. Territorial males in the breeding season respond to an invading call with what naturalists call 'a characteristic bonk vocalisation', a singularly apt description for part of the mating ritual. There are three distinct types of territorial call made by frog couples: one purely by males, another exclusively by females, and a third which they emit together. Some male tree-frogs have complicated calls which have two distinct components: one part is understood by females, another only by males. In this way they can serenade continually and indicate a wish to mate (which attracts females) coupled with a territorial warning the females ignore (but which repels other males).

Naturally, there is a downside to all this. Some small frogs are preyed upon by larger toads. The toads use their own calls to stimulate the little frogs into calling, and this call is used to track down the frog which the toad then proceeds to eat. Nature sometimes turns the tables when you least expect it.

How amphibians see

Frogs, salamanders and newts all have a good sense of vision. They use their eyes in many ways, much as we do: to find their way around, to return home, to select a suitable habitat and to find food. These animals do not see in the way we do. Their visual acuity is specialised, and different sight senses perform different functions connected with survival and feeding. Studies have revealed four different aspects of vision in the frog:

○ Detection of local sharp edges and contrast.

○ Movement of edges.

○ Local dimming produced by the frog's movement or by rapid general darkening (the approach of a large animal, for example).

○ Edge curvature of a nearby dark object.

The last of these has been called 'bug perception' because it is adapted for the identification of small insects on which the frog feeds. Experiments have shown that the greatest stimulus is sent to the brain when a small object (smaller than the field of view) enters that field of view, stops, and then starts moving again. This response does not depend on the quality of the background, and is not affected even if the background itself is filled with randomly moving objects. What we are saying is, of course, that frogs have excellent eyesight which is specialised for spotting their prey.

Toads, when they spot something to eat, first turn to face it.

If a creature is too small to be worth the effort, or too large to be caught, the toad ignores it. The response has been shown to be related to the contrast of the prey against the background, and also to the number of different moving organisms in the toad's field of view. Threatening movements of a large prey cause the toad to turn away, as though avoiding contact. At least, these are the scientific findings on toads confronted with a range of choices under experimental circumstances. One has to say that similar responses could be elicited in humans. If you were to be given the results in that terse and clinical fashion, you would be able to retort, 'What you are saying is that I have excellent vision which can distinguish between tasty food and a threatening enemy. Of course. I knew that!' Similar comments could apply to the frogs and toads. We can break down the components of vision and identify the specific responses associated with disparate objects in the field of view, but in the end we should accept that these amphibians have excellent eyesight for the lives they lead. In that sense, they are just like us.

Salamanders walk directly towards prey, and then freeze when within striking distance. As soon as the prey makes a further movement, the salamander strikes with its tongue and seizes the prey. Here too there is a temptation to list the stages as though they were sequences of automated responses. It is simpler (and may be far nearer the truth) to draw parallels with human behaviour. If you wish to consume live prey, then movement would attract you towards it. If the prey were stock still, you might well wait a moment to check it was still fresh, and alive. The clue would be further movement in the prey, at which sign you'd catch it if you could.

Look at it from the prey's point of view. This creature is skittering away as though nothing were amiss. Suddenly a monstrous animal is glimpsed looming nearby, so it freezes to check out the situation. In this way it will no longer attract so much attention, and the benefit of staying perfectly still is to make it much easier to spot whether that massive object really is approaching, after all. The seconds tick by, and the prey feels satisfied that the movement has ceased, so the threat must be over. It moves and, in a second, becomes lunch for the hungry salamander. In analysing these movements we must not forget that the prey are creatures, too, and creatures with sentience.

Amphibians have senses which we lack. Many experiments have revealed how they can orient themselves towards home if placed on a featureless platform with nothing but the sky above. It is known that they can navigate using the position of the sun in the sky, and some seem capable of navigating by the stars, or stars and moon combined. The sense organs which enable them to do this are not always the eyes, though we have yet to get to the bottom of the problem. It could be that polarised light is sensed by newts and salamanders, which would help them when they are out of direct sight of the sun (in woodland, for instance). Neither of these navigational aids is used by mere humans. Most intriguing are comparative observations on types of salamanders which are respectively sighted and naturally eye-less. These creatures can orient themselves with respect to polarised light, irrespective of whether they can see. There is obviously a means of sensing elsewhere on the body.

Amphibians certainly respond to different colours. Mature frogs, for instance, are drawn towards blue light. It has been deduced from this that the purpose of the response is to help them to find their way back to water. However, many types that do not go to water during their adult lives (even to breed, as is the case with *Eleutherodactylus*) still show a preference for the colour blue, so it is a widespread trait even where it no longer has any survival value. Tadpoles congregate more frequently around green objects, which is linked to their preference for water-weed. During metamorphosis from tadpole to frog, the shift of preference from green to blue can be observed. Sight is crucially important to frogs and toads, while salamanders and newts rely far more on their sense of smell and taste.

Amphibians and territoriality

Amphibians love home. They recognise where they come from, and usually return to the same place to breed. If displaced, they can find their way back over surprisingly large distances. They often mark out their own foraging territory and protect it with noisy displays or by aggression against an invader. Mostly these shows are for the benefit of other members of the same species, but in some cases territoriality is maintained by amphibians of one species communicating with those of another. *Plethodon* salamanders of North America identify territory and patrol it regularly. *Plethodon nettingi* can communicate with a different species, *P. cinereus*, and the two avoid each other's patch of land. This helps to cut down competition for limited food supplies.

In any given area, the number of different amphibian species is related to the moisture of the territory. Amphibians need to keep moist, and the less threatening their home base, the more tolerant they are of invaders. Moisture implies food, too, so they can afford to be more tolerant of other types sharing their patch. The species that live together tend to be those that do not compete with one another. Thus, one species may prefer larger prey than another; one may feed in the day, another at night; some may feed at the top of the leaf litter, while others prefer grubs found lower down. In each case, the creatures have finely attuned senses which enable them to identify where they wish to be, and also to sense what the others are up to. These communities depend on acute sensibility, and not on mathematical chance.

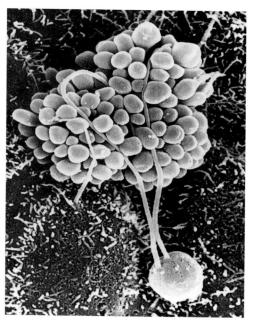

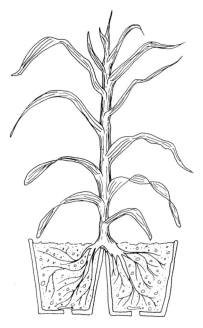

In some cases we are discovering new sense organs, but still know nothing about their purpose, or the senses they detect. In salmon, these strange bulbous structures in the brain seem intimately connected with sensory nerves. They appear to be some kind of sense organ new to science, perhaps detecting pressure. However, this remains speculation and there is little evidence of how they work.

Terry Parker, Nottingham University

A pot plant can be grown with the roots divided into two containers. After a period of normal watering, one of the two pots is left dry. The plant closes down the stomata in the leaves – the normal response to drought. Yet the plant is still receiving plenty of liquid from the watered pot. This simple experiment reveals the role of the roots as sense organs which send signals to the leaves.

Brian J. Ford (after Terence Mansfield FRS)

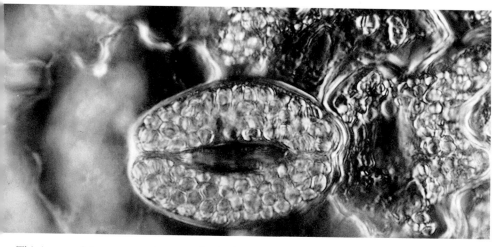

This is one of the stomata, tiny pores in the leaves of higher plants, which detect a range of stimuli. Each responds to the intensity and colour of light. Stomata are affected by the gases in the air (particularly carbon dioxide), by vibrations caused by touch or by wind, and by substances produced by insects on the leaf. They also have a sensory connection with the roots. In human terms, these are the senses of smell, touch, sight and taste.

Terence Mansfield FRS

Can trees change the position of their branches? Great storms have battered the British landscape and uprooted trees. This one, in Suffolk, was dislodged, but not demolished. Many trees will shed branches over time, but it is claimed that the major branches of this one have slowly shifted their position over the years until the body of the tree has regained its centre of gravity.
Bill Vinten, Suffolk

Plants have senses to interact with many other varieties of life, and we know that they respond to fungi, bacteria and insects. This scanning electron microscope image shows the roots of *Genlisea aurea*, which reveal characteristic chambers spiralling along each root. Specially adapted hairs allow ciliated protozoa to swim into the chambers, but prevent their escape. This is analogous to the insectivorous plants – but in this case we have the first discovery of a land plant trapping microbes for food.
W. Berthlott, University of Bonn

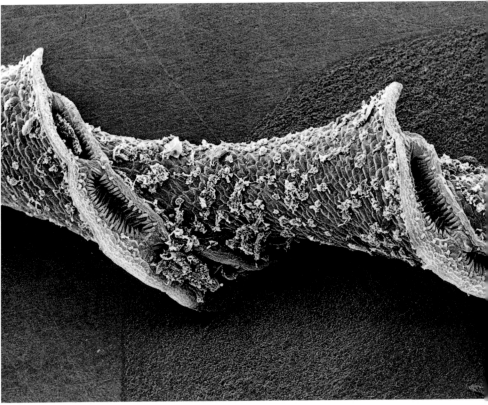

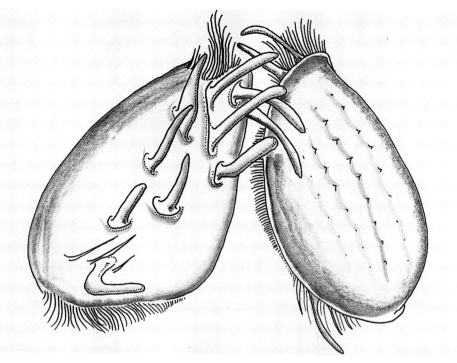

Ciliated organisms are complex cells with a sophisticated life story. In autumn they often undergo sexual reproduction. Here we see two *Euplotes* cells from a sample of pond-water in the act of caressing each other. This mutual recognition has recently been captured by the electron microscope. It is tempting to dismiss such observations as 'anthropomorphism', but we would do well to remember that *Euplotes* was doing this long before humans even evolved.
Brian J. Ford (with acknowledgements to Romano Dallai, University of Siena)

An adult chimpanzee (*Pan troglodytes*) selects leaves from the forest canopy in Sierra Leone. The chimpanzee has the ability to choose a varied diet from hundreds of species of plants and animals, including fruit, termites, birds' eggs and small mammals. It has a knowledge of medicinal plants, and can choose leaves to cure specific disorders.
Nick Gordon, Oxford Scientific Films

The worker bee cares fanatically for the young, works assiduously in the colony, yet can be a cannibal and a fierce warrior as the situation demands. Bees recognise their niche in the community, and carry out complex constructional and navigational tasks with unerring accuracy. If their honeycomb is damaged, they set out to identify and deal with the problem.
David Thompson, Oxford Scientific Films

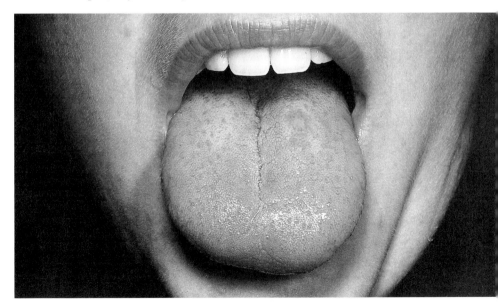

Our tongue is one of the most sensitive organs of the body. It is covered with projecting sensory papillae, each of which bears 200–300 separate taste buds. Salty and sour or acidic tastes are primarily detected at the sides of the tongue, sweet tastes at the tip. The sense of bitterness is concentrated in the circumvallate papillae at the back of the tongue, nearer the throat.
G. I. Bernard, Oxford Scientific Films

These African elephants (*Loxodonta africana*) exhibit complex social behaviour, showing great loyalty to each other and discomfiture at the suffering of others. Here they are at a salt-lick, picking up the sodium chloride they need to replace losses through perspiration. Western visitors to the tropics learn how to supplement their diet with extra salt, yet we can see that the elephant is well acquainted with the need.
Steve Turner, Oxford Scientific Films

The scarlet macaw (*Ara macao*) and the red and green macaw (*A. chloroptera*) collect scraps of minerals from a cliff face in Peru. Both species have complex community structures and consume a well-balanced diet. Grit helps to grind food within the muscular crop, and limestone can ensure that the bird does not run short of calcium during egg production. Soluble minerals serve to give trace elements necessary for health.
Günter Ziesler, Bruce Coleman Limited

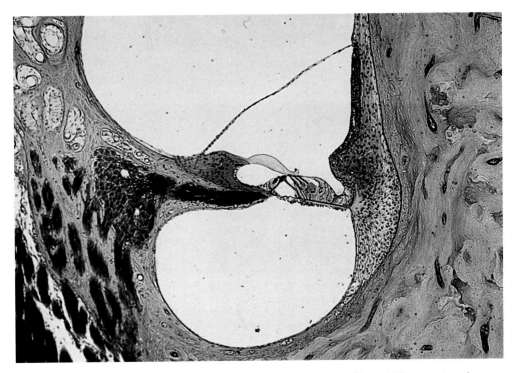

We detect sounds and the frequency of notes within the spiral cochlea which comprises the inner ear. This is the organ of Corti, stained for microscopical examination and thinly sectioned. Sounds resonating through the tapering chambers of the cochlea set the tiny hair-cells into movement and the resulting nerve transmissions are interpreted by the brain.
Oxford Scientific Films

Adult herons have been seen to drop scraps of bread into the water to attract fish. Herons wade into water from the sloping bank. This 'skating' grey heron (*Ardea cinera*) is confused by trying to step into water which has frozen overnight.
Manfred Danegger, NHPA

Fox-hunting: glorious tradition, or heinous inhumanity? Hunting the fox with hounds is a direct descendant of our prehistoric ways, and much tradition is lost when hunts are curtailed. Opponents point out that the resulting cruelty to the fox, no matter how similar to hunting in nature, seems inhumane when orchestrated by a refined human species.
Ronald Toms, Oxford Scientific Films

The common jay (*Garrulus glandarius*) thrusts worker ants into its plumage. Birds of the group known as passerines sometimes use ants to help preen their feathers. The birds select only those ants which secrete formic acid, which acts as an insecticide, and they always avoid the biting species, so clearly they can distinguish one species from another.
D. J. Saunders, Oxford Scientific Films

Young macaque monkeys of the Joshinetsu Plateau in Japan like to make snowballs. They gather snow by rolling it along the ground, just as children do. Although the adult macaques do not make snowballs themselves, they will play with those made by the young.
Mitsuaki Wago, Robert Harding Picture Library

The octopus has excellent vision and a brain so well developed it is often included in the same category as mammals in legislation governing animal welfare. The sense of taste of an octopus is at least 100 times better than ours, and they can hear as sounds frequencies far below our threshold.
Jane Burton, Bruce Coleman Limited

The venus fly-trap, *Dionaea muscipula*, was described by Charles Darwin as the 'most wonderful plant' in existence. Each lobe of the opened trap bears three fine sensory spines, and if one of these is stimulated the leaf will snap shut, imprisoning the fly. If the stimulus is too slight, the trap does not close; but if a further slight stimulus occurs within half a minute the plant recognises that a living insect is probably responsible and the trap closes anyway.
Sean Morris, Oxford Scientific Films

The honey-bee *Apis mellifera* gathers nectar for storage in the hive, where it is processed into honey. The bee uses the direction of the sun as a prime aid to navigation, and communicates the position of food sources to its fellows by a circular dance first documented by Karl von Frisch. The dance is performed in the darkness of the hive, and we now recognise that the attendant bees sense the directions of the dance by air vibrations, and not by sight. *N. A. Callow, NHPA*

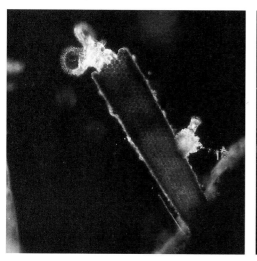

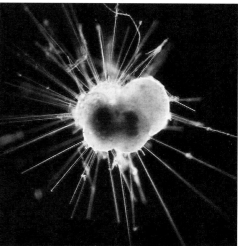

Rotifers are tiny organisms just visible to the naked eye and smaller than fleas. Each individual within a species has the same small number of cells per organ. This shell-building rotifer, *Floscularia ringens*, has slowly emerged from its shell to catch passing bacteria and algae which it will use as food.
Tony Saunders-Davies

A feature of the deep Atlantic Ocean is an ooze composed of foraminifera, a group of single-celled shell-forming organisms. Most are about 0.5 mm across, though some can be up to 20 cm in diameter. This *Globigerina*, with its fine spiny projections, is a common organism in plankton and often measures up to 2 mm in diameter.
Peter Parks, Oxford Scientific Films

Many plants, such as this male white bryony (*Bryona dioica*), climb with the aid of tendrils, which pull them up as they grow. Tendrils are modified leaves, in which the lamina of the leaf is lost and the veins develop an ability to twist and form a firm support. Once the tendril is holding firm, it twists like a telephone cord, shortening its overall length and drawing the plant away from the ground.
Jane Burton, Bruce Coleman Limited

The genus *Passiflora* is distinguished by its striking flowers. Each features a cross-like pistil, a ring of stamens like a crown of thorns, and about twelve petal-like sepals (the same number as the disciples of Jesus): hence the name passion flower. The tendrils will start curling towards the direction of a touch within thirty seconds of the stimulus.
Brian J. Ford

Reptiles have adapted to life on land through the development of special senses. This lizard, *Sphenodon*, has good eyesight and can accurately locate its next meal. The ears of reptiles vary greatly, but all species possess a sense of infra-red radiation. They use this to control their position in sunshine or shade, and can regulate their body temperature with remarkable accuracy.
Brian J. Ford

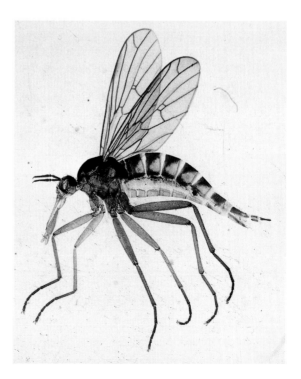

Mosquitos locate their prey by sensing the heat radiated from a mammalian body, and can also sense carbon dioxide exhaled by potential victims, lactic acid and moisture. Rounded sacs known as scolophores at the base of the feathery antennae can detect sounds, and the antennae themselves detect changes in temperature less than 1°C. Their compound eyes give them excellent vision in a field extending for 360° round the head.
Brian J. Ford

A young adult polar bear in an attitude of complete submission as it plays with a husky dog. The bear would make a meal of the dog in other circumstances, but in the spring sunshine the mutual hostility of the two enemies can be subsumed by play. The actions of the animals reveal a surprising level of mutually pleasurable interaction.
Norbert Rosing, National Geographic Image Collection

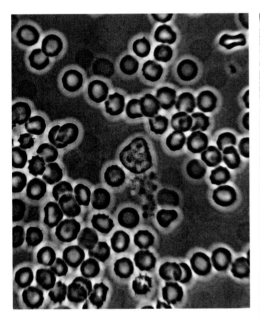

In this sample of living human blood, most of the cells are erythrocytes (oxygen-carrying red cells). At the centre of the picture is a group of minute platelets (concerned with coagulation) and a single leucocyte (white cell). The leucocytes use their senses to navigate around the body, seeking out invading bacteria and extirpating them. These cells are crucial for survival, but are not under direct human control.
Brian J. Ford

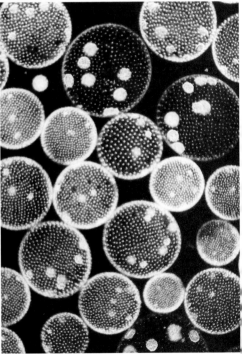

The green alga *Volvox* exists as spherical colonies of separate cells carrying paired flagella which beat steadily in unison, slowly rotating the spheres about their axes. The rate of beating is regulated by the entire colony so that the cell proceeds forward, rather than spinning on the spot.
Hilda Canter-Lund

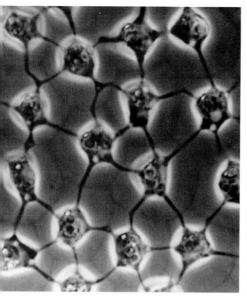

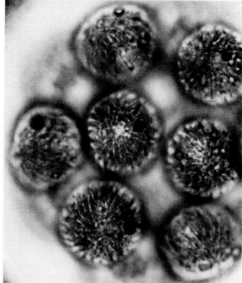

The secret of the *Volvox*: each separate cell is in contact with its neighbours through fine cytoplasmic branches. Each cell also has an eye. The rounded body within each cell (see in particular the central cell in this picture) is a light-sensitive orange-coloured pigment body which enables the colony to avoid extreme brightness while remaining in enough light to survive.
Hilda Canter-Lund

A high-power objective lens reveals the eye possessed by each *Eudorina* cell. The sense of sight is due to these dark organelles, each containing a light-sensitive pigment. By adjusting the direction in which each colony moves, the individual cells are able to convey some kind of image of their surroundings. Each eye-spot is no bigger than a typical bacterium.
Hilda Canter-Lund

Rounded cells of the alga *Eudorina* form spherical colonies. They line up in a symmetrical array and are one of the reasons that water in a pond turns green. *Eudorina* colonies contain fewer than thirty individual cells; the tiny colonies swim actively through pond-water and (like all green algae) capture solar energy for life through photosynthesis.
Hilda Canter-Lund

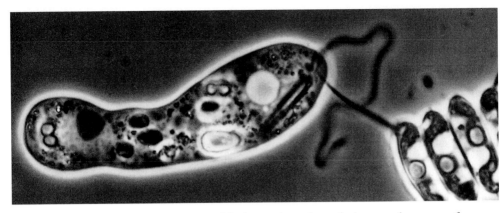

An inhabitant of still pools, *Peranemopsis* is drawn along through the water by means of a flexible flagellum (visible to the right). This is partly a means of movement, but also has a sensory function. Within the cell is a complex arrangement of food storage bodies and organelles. The nucleus is the larger grey body towards the left of the cell, and food granules and energy-releasing mitochondria can also been seen.
Hilda Canter-Lund

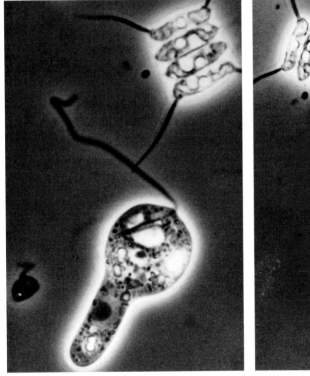

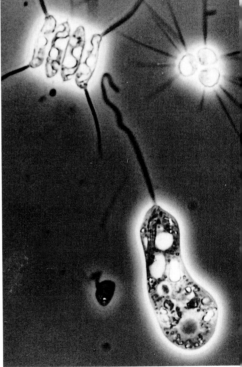

The *Peranemopsis* cell changes shape as it swims. This one is turning left to avoid four cells of the green alga *Scenedesmus*. These algae grow in groups of four, lying side by side, with four curving spines radiating from each cell group. Even within these tiny cells the complexity of organisation is visible to the optical microscopist.
Hilda Canter-Lund

In *Peranemopsis*, only the tip of the flagellum is active. It flickers through the water as it progresses, sensing the way ahead, and drawing the cell along behind it. *Scenedesmus* cells are also visible (above left) and to their right we can see a group of *Micractinium* cells, surrounded by fine radiating threads.
Hilda Canter-Lund

Among the many protists which make a home out of minerals are the tintinnids. These are ciliated microbes where the cilia are found at one end of a cylindrical or tapering cell. The organisms collect tiny objects from their surroundings and cement them together, producing a shell which protects them.
Hilda Canter-Lund

The eye is one of the most difficult of all organs to study under the microscope. In this section through a young eye, the most conspicuous structure is the lens. Curving towards it is the dark iris and in front we see the cornea. The rear of the eyeball is lined with the retina. This section does not pass through the optic nerve. The adult human eye contains a lens that is proportionately far smaller, and in land animals the cornea does most of the focusing of light.
Oxford Scientific Films

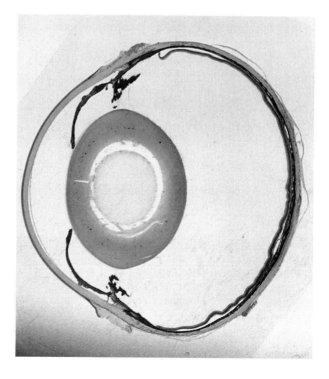

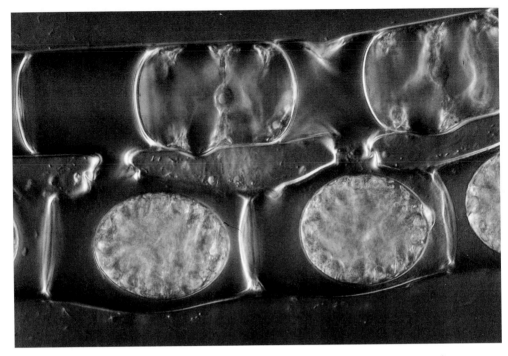

Two filaments of *Spirogyra* (the 'witches' hair' alga) undergoing sex. The filaments line up alongside each other and can sense maleness and femaleness. Side branches of the cells grow towards each other and fuse, producing a tube that joins the cells. The cells of the male filament (above) have migrated downwards to fuse with the female cells, producing dark, rounded zygotes. The two cells remaining show slight bulges where a tube started to form but was beaten by another. They remain unable to fertilise a partner.
M. I. Walker, NHPA

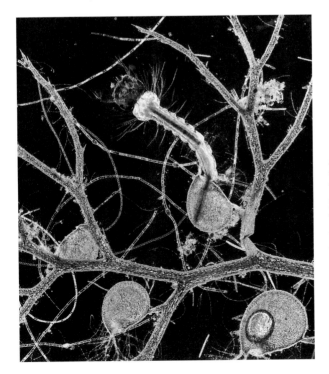

A silvery mosquito larva has its tail caught by *Utricularia*, the bladderwort, which traps passing animals – more usually water-fleas – and uses them as a source of food. Although these larvae are larger than the traps, they are sometimes caught. Their frantic wriggling eases them into the trap, curling up as they go, until the whole body is enclosed by the digestive organ of the plant.
Kim Taylor, Bruce Coleman Limited

Vision on dry land

Reptiles have left the water behind and adapted to life on land. This brings several changes in the way the eye operates. First is the reduction in importance of a lens in the eye. That is a surprising observation, I know, and I say it for several reasons. For amphibians, and other organisms living in the water, the effect of the habitat on light is very different from the effect of light on land. Amphibians and other small creatures are largely composed of water, and they spend most of their time in a watery environment, so they are accustomed to seeing underwater. The passage of light rays from water into the eye has little effect on the rays of light: they are simply being transmitted from one watery medium to another. The semi-crystalline lens of the eye is all that is needed to focus the image on the retina.

The sense of sight in land animals is very different. Light travelling through air, and then captured by the eye, is passing from one medium (air) to another very different, watery medium. As a result, the light rays are refracted severely when they first enter the eye. This is why land vertebrates do most of their focusing not with the lens as you might imagine, but with the cornea which covers the front of the eye. This change from water to air as the natural environment has several important effects on vision. First, the optical centre of the eye moves forward in a land-based animal, so that an eye of the same size will have a larger retinal image and will be able to resolve finer detail. Second, since water absorbs both red and blue light the spectrum is more limited, and land animals have a broader range of colours to see. And third, water scatters light, so underwater objects lack contrast compared to those seen on land. There are several important drawbacks to an eye which works in the open air. Ultraviolet radiation can damage the eye of a land creature in sunlight; and the problems of drying necessitate the development of tear glands, ducts and eyelids to protect and lubricate the surface of the eye.

Reptiles first evolved as animals of the daylight hours. They needed to be out and about during sunlight in order to warm their bodies and get up to speed. Their retinas were composed exclusively of cones, the cells which receive colour signals and give us colour vision. In time, mammals came to dominate the scene and with this increased competition, reptiles took a back seat. The diurnal forms (those active in daylight) became increasingly secretive and spent much time in hiding, while many others emerged only at dusk and became accustomed to a darker and more gloomy world. In time, reptiles adapted so that many of the cones in their eyes took on the appearance and function of rods, the cells that perceive contrast in black and white. It has been shown that the process can also work in the opposite direction. In animals which began life as nocturnal, and had a retina composed of rods, the change to a diurnal routine was accompanied by a modification of the rods so that they changed into cones and became sensitive to colour. Cells are basically adaptable, able to take on other functions should the need dictate.

The return to water

Just as reptiles emerged from the water onto dry land, some have returned to the water. Turtles and crocodiles are well-known examples, and they have adapted their sense of vision so that they can see under the water. Once submerged, the water-based cornea has little effect on rays of light entering the eye, so the animal relies on its lens to focus an image. These reptiles have an added problem. Because they evolved from land-dwelling organisms, most of them have to return to land to breed. Animals often have reminders of their ancestry in the way they breed. Amphibians like toads, living their adults lives on land, have to return to the water to reproduce. Turtles, having returned to the sea from a land-based environment, haul themselves back onto the land to lay their eggs in the sand.

On dry land, the cornea becomes the main focusing structure, with the lens performing a secondary role. Thus the lens has to perform two distinct roles, depending on whether the animal is above, or below, the surface of the water. The lenses of the turtles which have conquered both environments are amazingly pliable. In some freshwater turtles there is a sphincter muscle around the rim of the lens, which can contract and make the lens bulge into a highly refractive focusing element. These turtles can instantly focus on objects above the water, or below it; if you keep your eyes open while you are swimming you will quickly notice that the human eye fails to make that change. By contrast, crocodiles have good vision only on dry land, and cannot focus properly underwater.

Body temperature

Much effort is expended by reptiles in maintaining their body temperature within a functional range. Too hot, and the animal may die of heat-stroke; too cold, and it cannot move. Lizards hide away at night among rocks warmed by the sun so that they can keep warm enough to move in the morning. As the day begins they hug the ground, picking up heat. As the temperature heats up they run about on tiptoe, keeping above the hot ground, and hide in the shade if the heat becomes excessive. Their abilities to detect temperatures (and to make sensible corrections) are well adjusted and constitute a set of senses which is clearly highly developed. Humans suffer from sun-stroke more often than reptiles, for all our fine intellect.

Some animals possess infrared detectors which they use to locate their prey. We all have this sense: hold out your hand near a hot-plate or an iron and you can sense the heat radiating from its surface with your palm. Special infrared organs exist in pit vipers and rattlesnakes. When the organ develops in pit vipers it can be seen as a hole between the eye and the upper lip. At the base of this aperture is a highly sensitive membrane, rich in nerve-cells. Because this heat-sensing membrane is at the bottom of a pit with a small opening, a heat image forms in a similar way to the image formed by a pin-hole camera. The snake can see its prey in the infrared by using these organs. Experiments have shown that these snakes can strike their prey even with their eyes covered. What is more, they could catch prey even if it moved after the snake had begun its strike. These creatures have a sense we all know, but it is developed to a high degree. We warm our hands by a glowing fire; they can follow their prey in the dark.

The third eye

Deep within the brain lies a small gland, the pineal body. In humans it is a gland that is connected to the eye. During darkness, at night, this gland secretes melatonin. This secretion activity ceases during the day. The melatonin regulates mood and has effects on the sex organs and the thyroid gland, among others. In animals it has been found that decreases in melatonin can prolong the breeding season, and it is thought that melatonin levels can be related to puberty in human beings. The pineal gland may well regulate hibernation in mammals. Where did the pineal body originate? It is a strange little gland, reduced to only a few cells in some giant mammals (whales and elephants have hardly any pineal body at all). We still know little about its functions, though it does have this subtle connection with the eyes.

In other animals, the pineal body is easier to understand. It is a third eye. There is no mistaking its function: it has a lens, a retina and an optic nerve. In the lamprey it is prominent, and looks like a small eye in the middle of the head. In other animals, including some reptiles and birds, it still has the ability to detect light. In lizards the middle eye lies beneath a hole in the skull, and it peers out through a pit in the skin. Within the retina are rods and cones, as in any normal eye, and there is a nerve from the retina to the diencephalon of the brain. This central eye seems to respond to overall light and heat levels. Lizards without one tend to be less thermally adapted than those that have a functioning pineal eye. It also affects the thyroid, the activity of which seems to be dampened down by the activity of the middle eye. Interestingly, the middle eye is more developed in lizards which live at higher altitudes than in those living near sea level. In the lizards it is now believed that this third eye may be sensitive to the coloration of the sky at dawn and dusk, so it may serve to regulate the daily cycle of activity.

Whatever the full story of this strange little organ, there are animals which have an accessory eye and to them it is clearly important. Humans, with their tiny pea-sized pineal body regulating the mid-winter blues, lack this extraordinary 'second sight' which some of our distant relatives still possess.

Do fish have emotions?

As cold-blooded, heartless automatons, fish bludgeon their way through life, paying little attention to the outside world. Or do they? I have often dived through the shoals of silverside fish in the tropical oceans. These beautiful fish respond like one living organism. The thousands of individuals in a single shoal turn and twist, passing instantaneous messages to each other and responding at almost lightning speed. They have a finely tuned set of senses which allows the entire community to move as a single entity. Sometimes we see flocks of birds doing this, but fish are masters of the art. Recent research has thrown light on the extraordinarily highly developed senses with which fish are endowed. It also shows that even that seemingly unattractive sea-creature, the dogfish, can pine for its freedom and resent imprisonment. When caught and left in a tank, dogfish start to deteriorate. They are so traumatised by the experience that they will not eat. At first they lie still, sometimes for several days. Some – without obvious signs of injury – even die during this time. Then suddenly the moping dogfish come back to life. After lying still and unresponsive, they start to swim vigorously about the tank. For many days they swim ceaselessly, searching every corner and moving briskly around their new environment. They fight, too, particularly at night.

But in time, this behaviour starts to change. The frenzy of swimming is punctuated by periods of inactivity at the bottom of the tank. The fish lie still, before swimming actively once again. Then there is a further period of stillness. As time passes, the motionless phases grow longer until the fish cease swimming altogether. This time they are left lying listlessly or struggling weakly, and in a while they turn belly-up and die. Scientific investigation shows that these behavioural changes are related to alterations in blood clotting. During the phase of weak inactivity, blood clotting is slow and ineffective; in the swimming phase, by contrast, blood clotting is efficient and rapid. Confined dogfish cease feeding. Only in large tanks at oceanaria can captive dogfish be encouraged to eat. There are many other changes within confined dogfish which scientific examination can reveal. The cells of the pancreas steadily deteriorate. There is a fall in serum glucose and blood cholesterol levels. The structure of the kidneys changes, and rising levels of sodium, potassium and chloride in the blood lead to generalised dysfunction in all the organs. Muscle tissue throughout the body deteriorates and wastes away.

These findings reveal that the dogfish, used for decades as a teaching subject for students of biology and medicine, needs to be reappraised. The captive dogfish is different in many respects from the creature swimming free in the wild. It also suggests to us that fish can resent being captured and pine for their freedom. The sequence of responses – lying still, then a phase of furious activity, followed by longer phases of inertia until death supervenes – parallels the behaviour shown in mammals, and even in humankind.

Fish behaviour

The atmosphere in which a fish exists is water. It is in constant change, and much of it is in ceaseless movement. Fish have specially adapted senses which fit them for the constraints of this existence. Their sight is good, and so is their sense of smell; but they also have organs to detect changes in water-pressure, and these can tell fish much about distant movements of other moving creatures, whether predator or prey. The eyes in fish are large, and the lens takes up the bulk of the volume of the eyeball. Sunlight does not penetrate deeper in the oceans than 200 m (about 650 ft), and most fish are found in this upper layer. The abyssal species which live at greater depths include some that produce their own light from phosphorescent organs in the skin, and this can act as a lure for their prey.

Fish have remarkable navigational abilities. Because of the changing nature of their environment, most fish migrate. Even those that live in an enclosed body of water like a lake may breed in one part and feed in another. Many sea-fish return to fresh water to lay their eggs. They are particularly adept at finding their way home over vast distances. Only one group of fish, the eel *Anguilla*, turns the conventional pattern on its head, for they breed in sea-water and spend much of their adult lives in fresh-water rivers and lakes. Eels can cut across fields if they need to, as part of their migration, and can travel for many kilometres on land just like snakes, so long as there is

moist vegetation to protect them from drying out. They have
enough of a sense of direction to carry them from the tropical
Atlantic to the same rivers their parents inhabited.

The traditional views of fish behaviour have downgraded
their abilities to the status of automatic actions brought on by
environmental expediency: they drift in currents, home in on
changes of temperature as a result of instinct, and so forth. Fish
are cleverer than this. After hatching, the young fish-fry explore
their immediate environment and make repeated journeys to
sites as they become familiar. The young fish venture farther and
farther afield, often accompanying adults on feeding expeditions. They do not follow predictable pathways, but actively
explore, much as we find our way around a new environment.
This is how salmon and trout return to rivers which were, until
recently, ecologically inhospitable. The Thames, as an example,
has been barren of migratory fish for a century but in recent
years has re-acquired populations of salmon and trout as bio-
remediation has purified the water. Returning salmon once
found the Thames unsuitable for colonisation and thus turned
away. Now that it is of higher water quality, these sentient and
magnificent creatures elect to use it once again. Their colonisation of the Thames results from exploration and inquisitiveness,
and not from a blind homing instinct that draws them inexorably back to where they were born.

In the late twentieth century, a rise in temperature of the
northern Atlantic has led to the establishment of completely
new migration routes for a number of species including the
Atlantic cod and the bluefin tuna. Rather than slavishly following inborn patterns of migration, these fish also seek to develop
new lines of contact whenever they can. Until 1920 the northern extent of the Mediterranean-spawning tuna was the western
extreme of the English Channel; now they are feeding off the
coast of Norway. Before the Second World War the Atlantic

cod maintained a breeding colony off the southern tip of Iceland. They now regularly migrate around the south coast of Greenland to extend up the far west coastal waters, solely because of the cod's natural tendency to explore and to investigate new territories.

Fish as parents

Some fish go to great lengths to lay their eggs and protect their young. Even carp, normally considered to be large and lazy fish, will jump small obstacles placed in their way as they explore upriver. Most freshwater species swim upstream to lay their eggs. There the water is shallow, which reduces the likelihood of predation by larger fish, and the fast-moving water is rich in oxygen. These higher reaches are lower in food, so the young fish-fry are equipped with a sizeable yolk-sac to sustain them as they drift downstream after hatching to areas where food supplies are richer. Salmon struggle to the highest reaches of a stream before laying and fertilising their eggs. They use their fins to wash away sand and small gravel from the side of a rock on the river-bed, and lay the eggs ready for fertilisation in the resulting trench. The fish know exactly where to do this, for the currents that eddy round the rocks bring a layer of sand back again to cover the eggs several centimetres deep. The young fish have to struggle through the protective layer to escape downstream, but are able to mature out of sight of would-be predators. The salmon expend large amounts of stored body energy in the process. Fertile salmon, ripe with eggs and spermatozoa, make repeated leaps up raging torrents and steep waterfalls as they ascend the river. They do not feed on the way upstream to the fresh-water breeding grounds, and the majority of them do not survive the return journey.

Many other fish lay their eggs under cover in this way. One of the most spectacular examples must be the grunion, *Leurethes tenuis*, of southern California. The grunion is a smaller relative of the mullet, and comes onto the beach to spawn. Its breeding is synchronised with the spring equinox in March, when the tides are at their highest. The highest tides are at full and new moon, and the grunion selects the three or four days of the highest tides to swim *en masse* towards the shore. The first fish to ride the waves are males, and they are followed within half an hour by huge schools of females. The grunions time their run to coincide with waves which run up the shore. They are carried up onto the sand by a wave. Each female wriggles furiously to excavate a trench several centimetres deep, in the bottom of which she deposits about two thousand eggs. Males in attendance fertilise them. The entire process takes less than half a minute. Then, when the next sizeable wave approaches, the fish strike out for deeper waters and are swept back to sea. Timetables are published locally so that enthusiasts can come to watch the astonishing sight of teeming masses of silver fish wriggling frantically on the beaches.

Jawfish live in an excavated vertical tunnel which they line with shells and small pebbles. They use this as a base from which to launch raids on passing prey. Considerable care is taken to select suitable reinforcement materials, and running repairs are carried out as necessary. This remarkably coordinated behaviour should not surprise us unduly, bearing in mind that insects (like caddis-fly larvae) can construct homes, and so can single-celled amoebae. Among the fish that build seasonal nests is the sand goby, which uses a shell to construct the roof of its shelter. It finds a suitable shell (a scallop, for example) and positions it with its fins. From beneath, it systematically excavates enough sand to make a large cavity, which it seals with a layer of sticky mucus. The fish swim upside-down to lay their eggs, which are glued to the inside of the shell roof by more mucus. The whole nest is well camouflaged by this method of construction, and if the nest is damaged from time to time the male carefully repairs it. To do this he clearly uses his own mental abilities to assess the problem and rectify it.

The stickleback *Gasterosteus* constructs a nest from water-weed. In early spring the male stickleback stakes out his territory. His belly turns bright red, which acts as a sexual attractant. The male sets about building by gathering fragments of aquatic vegetation, which he glues together with a secretion of the kidneys which hardens underwater. When the nest is ready, he dances in a courtship ritual to lure a female to it. Female sticklebacks undeniably find the red belly attractive, but (given two males of similar appeal) they opt for the bravest male: that is, the male which swims closest to would-be predators as he frightens them off. The male nuzzles his partner to induce her to lay eggs. Once she has done so, he fertilises the eggs and then chases her off. He repeats the process, until several females have laid their eggs in his nest. The male then devotes himself to fatherhood. He fiercely protects the nest from would-be predators, and wafts currents of fresh water through it with his tail. He watches the eggs as they hatch, and keeps predators away from the young fish-fry until they are sufficiently mature to leave the nest. This sequence of events parallels the care for the young shown by mammals and by many birds. Nest-building and protection are concepts which imply decision-making, especially since no two stickleback nests can ever be alike. These adult fish are devoted to their task, and we need not invoke spurious and simplistic mechanisms simply to avoid expressing wonder at what we see.

Senses of fish

The sea, which seems so silent and calm once we sink below the surface, is actually a noisy place. Not only do the oceans ring with the eerie cry of whales and their kin, but the very noise of the waves on the surface travels for great distances. There are other sounds, too, which we are only just beginning to recognise. Fish talk. They make clicks or grunting sounds, and have a language for interpersonal communication which can be detected by purpose-built microphones. In the tabloid newspaper the *Sun*, this research was headlined in a bright and breezy fashion: 'Randy Haddock revs up for Love'. In science we often complain about the vulgar way in which biology is portrayed in the popular media, but it is hard to fault this description. The sound made by the sexually aroused male haddock sounds exactly like a motor-cycle revving up at the start of a race.

Recordings of the haddock show a lengthy litany of call sounds. The fish emit occasional clicks, but the frequency of these can increase until they are being produced at the rate of 300 Hz. The cause seems to be muscular contraction, but our muscles cannot emulate this rate of response. If normal muscles contract faster than 30 Hz, they seize up. They accumulate metabolic products and are thrown into spasm. The strange mechanisms which allow fish to circumvent these restrictions are yet to be understood.

Some fish utilise the swim bladder to create sound. This is an air-filled sac with which the fish control their buoyancy. Some fish have them; others don't. In some species, sounds are produced by muscles alongside the swim bladder. Other sounds are created by rubbing hard parts of the body together, so that a creaking or rasping noise is produced. In many fish, a range of sounds is emitted in different social circumstances, and the fish clearly communicate as they maintain their communities.

Training fish to respond to a given sound is an easy matter, for they can be taught to associate the noise with the release of food. This gives us a way to test the importance of the swim bladder as an acoustic organ. At the Marine Laboratory of the Scottish Office of Agriculture a series of intriguing experiments shows how the system works. The swim bladder of the cod can be emptied with a syringe. If this is done, so that the fish lacks the acoustic assistance of a gas-filled sac, it loses its hearing. The fish becomes partially deafened by the loss of the pouch of gas within the body. Over the following hours, the fish rein-flates its swim bladder with gas liberated from the bloodstream. As the sac inflates, the hearing returns to normal. The fish seems to be using the resonating air as a means of amplifying sound. What of fish which naturally lack a swim bladder? Are they at an acoustic disadvantage? Perhaps they are. The research team inserted a fine plastic balloon into the body cavity of plaice, and gently inflated it with air. Plaice lack a swim bladder, but in this experiment they were gaining one for the first time. Tests of their ability to hear showed that the per-ception of sound was improved as long as the artificial air-sac was present.

Low-frequency sounds are widely used by fish to detect direc-tion and distance. There is a great vocabulary of sounds to which they can respond, and the human observer, listening to the tapes, can easily recognise the rising excitement or the sense of threat in the noises made by the fish. Here too one senses the universality of communication. We may not yet understand the language of fish, but we cannot mistake the mood.

Many migratory species of fish are able to navigate by olfac-tion, the sense of detecting odours and tastes in the water. Their sense of smell is far more developed than that of humans. Fish also possess an extra sense unknown to us, for they can detect minute changes in water pressure. Along the sides of fish are lat-eral lines, running from head to tail on either side of the body. Buried in these lines is a series of minute fluid-filled pits, and each of these is connected to a fluid-filled vessel which runs beneath the skin. Within this elongated, thin cavity is a line of sensors, the neuromast system, which creates a sensory potential and transmits signals to a nerve cord beneath the lateral line. In this way the fish constructs a mental map of the surrounding changes in water pressure. By detecting these minute alterations, created as objects swim past, the fish is apparently able to create a mental picture of solid objects by integrating the pattern of pressure changes each creates.

Fish can also detect electrical signals set up by the muscular activity of other creatures in the water. All fish produce minute electrical disturbances, for the simple reason that all muscular and nervous activity has an electrical component and water is a conductor of electricity. Skates and rays have sense organs in the skin, the ampullae of Lorenzini, which are sensitive to electrical charges in the water. Using these senses, fish can construct a three-dimensional vision of the other creatures in their surroundings in a way we cannot comprehend.

Some species of fish generate electricity of considerable power, and have been known for this unusual propensity since ancient times. The shocks caused by electric rays were written about by the ancient Greeks, though they believed that the reactions of fishermen who came into contact with the fish were caused by a poisonous effluvium which affected the blood. The electric catfish was known to the Arab and Persian philosophers of old. The ancient physicians used these fish as agents of electrotherapy, pressing them against the skin of victims of aches and pains, often with beneficial effects. By the time electrotherapy became popular in Western medicine during the Victorian era, it had been in use for thousands of years in the Middle East. The electric eels of South America have a mythology of their own, and are popular objects of mystique in books. These species can generate more than 600 volts, which is easily fatal to a human. The electrical charge is produced by layers of plate-like structures in each electrical organ. The plates produce no more than 0.1 volts each, but the extensive arrays

combine their effects in a natural battery to produce the synchronised output of a massive, potentially fatal, discharge.

The mudskipper *Periophthalmus* spends much of its life in
the open air, hunting for food on the surface of mud-flats.
These fish are found in estuaries where mangroves grow, and
follow the tide in and out, keeping mainly out of the water. At
high tide they climb among the roots of the mangroves, and
they sleep in hollowed-out roosts they make in the mud. These
fish have a well-developed sense of direction, for they can find
their nests even if the tide has covered them when night-time
falls.

A mental map of its surroundings is created by the goby
Bathygobius soporator which is found in tidal pools. When the
tide is in, it swims around its territory and explores the area,
making a mental map of each location. We know this because,
when the tide recedes and the area is transformed into a series of
tidal pools, these fish can jump from one pool to the next. Until
they have leaped from the water, no sight of the destination
pool is possible, so we know that the goby retains a map of the
pools in its mind. They hop from one pool to the next at will,
basing their navigation on the familiarity gained from purposeful exploration when the tide was high.

It is often said that fish have an attention span of only a few
seconds. Those who keep them say that fish can recognise individual humans, and they certainly know when feeding time
approaches. Can fish make judgements, based on a one-off
experience? Events at a neighbour's fish-pond may illuminate
this possibility. The pond was full of fish. Early one morning, as
shafts of sun sloped down past the lilies and waterweed, a heron
stopped by for breakfast. Its take-off was so noisy, flapping in
the still dawn air, that our neighbours were woken; when they
examined the pond they found it to be utterly bereft of any sign
of fish. A week later one of the children came in from school,

delighted that the parents had bought a fish to replace those taken by the heron. They hadn't. Over the following week other fish emerged, until in the end it was clear that the population of the pond was much the same as it had been before the heron's visit. The fish had simply been lying low. They seem to have remembered the heron (taking appropriate avoiding action) for weeks, not a few seconds.

A pet marine porcupine fish is said to have some awareness of time, and knows to within an hour when feeding time is due. The change of feeding from 8 pm to 11 pm reportedly made the fish adjust its reactions accordingly. The fish signals by shooting a jet of water out of the tank when it gets frustrated, and it is said to be able to distinguish members of the family from strangers. There are many anecdotal accounts like this, and we truly need some extensive research to elucidate the ingenuity of fish, rather than to confirm their supposed stupidity.

Pet fish pine for a mate. Fish left alone in a tank after a partner dies have even been known to leap from the water and die on the carpet. We know that fish can be self-sacrificial in their desire to breed, that gobies know where the next pool lies, and that goldfish can be trained. If they are given food at different times at the ends of the arms of an experimental maze, they can soon learn what to do and when to do it. Fish possess several senses unknown to humans, notably the pressure-sensors of the lateral line, an ability to visualise through electrical charges in the water, and the senses of the middle eye of lampreys. To regard fish as cold-blooded creatures may be definitively accurate, but we should understand their essential nature and recognise that all these creatures have abilities which warrant our interest, and feelings we should respect.

5

Plants Have Senses

Plants are bursting with movement. They are rich in sensation, and respond to the stimulation of the surrounding world every moment of their active lives. They can send messages to one another about overcrowding or a threatened attack by a new pest. Within each plant there is ceaseless activity as purposive as that in an animal. Many of them share hormones that are remarkably similar to our own. Their senses are sophisticated: some can detect the lightest touch (better than the sensitivity of the human fingertips), and they all have a sense of vision.

Trees manage to grow in well-spaced patterns, as a walk through woodland will confirm. They employ mechanisms designed to prevent overcrowding, which would lead to competition for food, light and water. Not only can plants communicate an attack by pests to other plants in the neighbourhood, but they can react to disease by chemical responses which parallel some of those seen in animals. Plants have great regenerative powers, and the way they heal themselves shows immense coordination of cellular growth. A tree from which a branch has been cut covers the site with wound tissue and makes good the damage. If you do not cut down a branch, then the tree may well do that for itself.

Trees have the ability to configure their outline during their lifetime. For example, they can shed their branches to maintain their equilibrium. An even more remarkable ability is reported by Bill Vinten in Suffolk, who reports that a tree which was partly dislodged by a gale has altered its branches to regain its balance. He observed that the tree had been left leaning downwind after the storm. Over the following years, no branches were lost from the tree, but those that remained have grown round to restore the tree's centre of gravity. We have no knowledge of how a tree does this, and the maintenance of the outline of a tree is clearly a result of its sensory awareness and is worthy of further study.

Plants have much in common with animals. The essential difference is that a green plant can capture sunlight and use its energy to power its life processes. The light from the sun is used by the plant cells to do something science cannot imitate: they take molecules of carbon dioxide and water, and fit them together to make carbohydrates. As long as plants are in the light, they will keep doing this. The simplest carbohydrate molecules are sugars, but among the more complicated carbohydrates is cellulose. Cellulose does not dissolve in water, as sugars do, so it has to be laid down where it cannot harm living cells. As a rule, the cellulose is deposited around the outside of plant cells. This means that each plant cell is surrounded by a cellulose wall, and is trapped inside this insoluble box. Whereas typical animal cells can stretch and change shape, can expand and contract and can easily divide into two, mature plant cells are surrounded by a stiff capsule of cellulose and are far less mobile.

The growing tip of a plant is made of thin-walled cells which can still divide. The direction of growth is normally towards light (plants can, to that extent, 'see') and upwards, away from the ground (they can sense gravity). As the cells mature they lay down their cell walls. Cellulose is the typical deposit in plant cell walls, though some plants produce other substances. Lignin, for instance, is widely found in woody tissues. Structures made of silica glass are found in some cell walls, which is why a blade of grass can cut the skin like a saw. It is the wall of a plant cell which prevents such plants from moving around. They are rooted to the spot, taking in water that evaporates from the leaves. This continuous current of water passing up along the stem carries nutriment through the body of the plant. The movement of water is called transpiration, and it is vital for a plant's growth.

The evaporation of water from a plant is regulated by pores (stomata) which can open and close as the plant dictates. They also control the passage of air and carbon dioxide through the network of cells within the leaf. Each stoma is controlled by two guard cells which can open and close the aperture between them. The chemistry is complex, but when the guard cells enlarge, so that the pore opens, potassium chloride migrates into the cells and starch breaks down. When the processes reverse, and potassium chloride is released from the cell as starch reforms, the guard cells lose volume and the pore closes. The guard cells do not simply control the pores, but sense what is going on around the plant. First, they react to the intensity and the quality of light. Second, the stomata can detect the chemical nature of the atmosphere, responding to levels of carbon dioxide and other gases. Third is their ability to respond to physical stimuli that affect the leaf, like vibrations and movement caused by wind, and their fourth sense is of substances produced by organisms on the leaf surface. As Terry Mansfield of the University of Lancaster has pointed out, these correspond to four of the classic senses: sight, smell, touch and taste.

Although green plants have no nervous system, they can transmit messages through the length of the plant body. It has long been thought that the stomata open and close solely in response to what they sense, but we now know that they can also be controlled from as far away as the tips of the roots. During periods of drought, the water flow up a plant diminishes and the stomata close down to minimise water loss. It has long been believed that it was the closure of these pores in the leaf which reduced the water flow. In the 1980s it was discovered that the stomata start to close down the moment the roots detect dry soil, and long before there is any change in the water reaching the leaves. The plants are anticipating a threat before it arises. The mechanism seems to be some form of chemical signal which the plants can send to the leaves, and which leads to the closure of the stomata *before* the plant experiences water loss.

One of the simplest methods of demonstrating this effect is the split-root experiment. A plant is induced to share its root system between two pots, which can be independently supplied with water. If the soil in one of the pots dries out, the stomata over the whole plant tend to close. This occurs even if there is a plentiful supply of water to the second pot. The crucial role of the roots can be demonstrated by watering the dry soil, for the stomata immediately open. Confirmation is obtained by cutting off the roots to the dry pot. As soon as the roots are detached, the signalling system is severed and the stomata open. The control of the stomata by the roots may be by means of abscisic acid, a hormone which can cause stomata to close if present in very small amounts. One part of abscisic acid in a billion parts of water is enough to make them shut. Analysis of roots from split-root plant experiments substantiates the possibility, for there is always much more abscisic acid in the dry roots than in the moist ones.

Plants are subject to a great range of stimuli, and have finely tuned senses to optimise their behaviour. Inside a plant there is ceaseless activity. Microscopic particles within each cell are moving around to catch the light to the best advantage, and cytoplasm inside the cells is streaming from one place to another. Water is drawn up through the stem, and elaborated food-stuffs pass down like a nourishing blood supply. A plant in woodland may seem static to the casual observer, but inside it is a hive of activity.

The man-eating plant

Plants that eat humans are known to most of us. There is one that emits a musical sound so enticing that nobody can resist it. As people come closer, the plant snaps shut and consumes them alive. Some are normally harmless, but can eat people whole when the season is right. Other varieties maim people, but don't actually eat them.

These are all fictitious species, of course. The triffid is the best-known example. This plant, featured in John Wyndham's novel *The Day of the Triffids* written in 1951, could strike someone blind and the end result was usually fatal. In 1955 James Schmitz published a story featuring a plant known as Grandpa. Normally, Grandpa grew like a floating lily-pad, but once in a while it was commandeered by a yellow-headed frog, and in this state it would eat anything. Its carnivorous tendencies could only be stopped by removing the frog. The melodious man-eating plant was Black Mouth. It was among the carnivorous vegetable monsters created by Brian Aldiss, who featured them in his 1962 book *Hothouse*. Human victims were attracted by a haunting song (resonances here of the sirens of Greek mythology, who lured sailors onto the rocks) while the trapping and eating element was based on a real plant, one which has inspired science-fiction writers for generations. This is the Venus fly-trap *Dionaea*, described by Charles Darwin as 'the most wonderful plant in the world'. We may have seen it on television, watched it in action in the video library of a CD-ROM encyclopaedia, or read about it in books. The species comes exclusively from the Carolinas of North America, where it is fast disappearing, but it is widely grown for its novelty value. As I write it is on special offer at the garden centre tacked onto a local supermarket. Most of us could now own this 'most wonderful plant' if we wished.

The fly-trap eats small arthropods like insects. If a scurrying creature stumbles across an open fly-trap it can be caught as the jaws swing shut. One of the CD-ROM encyclopaedias has a video clip showing a plant ensnaring a small frog, but the frog is clearly being pushed into the open trap by the camera-operator's assistant and it does not give a good indication of what happens in the wild. This plant has intrigued observers since the first botanists went exploring in the New World. It seems to be the closest to an animal, for it has jaws which snap shut in one-third of a second, and teeth which hold the prey. Darwin found it enormously intriguing. He observed it at length, studying how it trapped small creatures, and artificially feeding it with small fragments of meat and cheese. He was fascinated by the idea that the motion resembled that of an animal's nervous response.

Each trap is composed of a leaf divided into two oval portions, which open like a butterfly's wings. On the glandular inner surface of each lobe are three fine spiny hairs, and around the far edge are interlocking teeth like those of a man-trap. A slight movement against one of the hairs triggers the trap into action: the two halves spring towards each other so fast that the prey cannot escape. As the trap begins to close, the teeth on each side are already interlocking, so that the most agile insect finds its way barred. Once the creature is ensnared, the glands on the leaves start to secrete digestive enzymes. These break down the body of the prey into a soft, moist mass which can be absorbed through the leaves and nourish the growing plant.

Was the effect purely mechanical, or might there be a nervous response within the plant? Darwin knew that electrical impulses were sent along nerves. That had been discovered in 1843 by Emil du Bois-Reymond (1818–96), an eminent German physiologist. It would be a simple matter to use the same kind of recording apparatus for research into the fly-trap. Darwin lacked the expertise, and so in 1873 he sent some samples of the plant to a prominent physiologist, Sir John Burdon-Sanderson (1828–1905), who was working on animal nerves at University College, London. Burdon-Sanderson hitched the leaves up to the apparatus that he was using to study how nerves send signals to muscles, and soon found an electrical signal. The electrical impulse was recorded immediately the sensitive hairs were touched, and before any movement was detected. Here was a series of three events: stimulation of the hair, electrical impulse, closure of the trap. This is very similar to the way an animal responds to a stimulus. However, it would have been premature to conclude that the fly-trap was showing truly nervous activity. The electrical peak which Burdon-Sanderson recorded was certainly an action potential, like that of an animal nerve, because it was a peak of electrical energy, followed by a movement. But was the electrical activity the cause of the movement, or was it the result of the chemical changes going on inside the cells?

Chemical reactions are accompanied by changes in electrical activity. The positively charged nucleus of an atom and the negatively charged electrons that surround it remind us that electricity is at the heart of matter, and as atoms and molecules react, and ions form (these are atoms with extra, or missing, electrons), we can measure the creation of electrical activity. If chemical change were produced by the stimulation of one of the trigger hairs on a fly-trap, we would expect it to be detectable as an electrical response. Just because we measure a peak of electrical activity does not necessarily mean that there is a nervous reaction; we could merely be measuring the results of a chemical change inside the cells. This objection was raised at the time by Julius von Sachs (1832–97), an eminent German botanist, who concluded that there was no true nervous activity in the fly-trap. He had two main reasons for drawing this conclusion. First, the speed of the impulse was too slow. Impulses can travel along animal nerves at thousands of centimetres per second, while the rate in the Venus fly-trap was only 20 cm (8 in) per second. And second, there are no 'nerve' cells in the fly-trap. Surely, von Sachs argued, if there were nerve impulses, they must have nerves along which the impulses travel. This scepticism resulted in a lack of further interest in the study of plant movement, and the subject lay dormant for a century. But in the 1960s, US scientists started to study how impulses are transmitted by cells, and similarities between plants and animals started to emerge.

Life on earth first developed in the salt water of the seas, and one of the fundamental mechanisms of any living cell is a way of controlling the level of salt – sodium chloride. Keeping the sodium ions from salt water at bay is done by restricting their movement in and out of a living cell. The cell is covered with a thin skin, the cell membrane, which is a fatty layer insulating the inside of the cell from the outside environment. Because of the ions inside the cell there is a negative electrical charge (about one-tenth of a volt). The central layer of fat in the cell membrane acts as an electrical insulator. Ions can pass this barrier only if tiny apertures open to allow them through, and these gates are used by the cell to regulate the passage of ions through the membrane.

If an impulse passes along a nerve in an animal, an advancing wave of gates opens to allow sodium ions into the nerve and potassium ions out. Now that we can study plant cells, the same mechanism has been detected in them. The most interesting discovery of all was that plant cells, like animal nerves, can manifest a receptor potential before the action potential itself. The action potential is the electrical signal that induces a response in a living cell; the receptor potential is the signal created before that – when the stimulus itself is detected. The receptor potential results from the hair-cells in the fly-trap being touched. This stimulus is translated into an action potential only if it is strong enough.

So a small stimulus, producing a modest receptor potential, may be insufficient to cause the action potential to be triggered; in which case, the trap stays open. A strong stimulus will create a receptor potential sufficient to generate an action potential, which closes the trap. But what happens with medium-strength stimuli? If a stimulus is below the threshold, the trap remains open; but if several more small stimuli are received, the hair-cell will still generate an action potential. It is as if the cell is remembering the stimuli and adding them up, which suggests that there is a kind of memory within the plant cell.

This means that a tiny fly will not ordinarily be trapped, but a slowly moving insect, which produces only the slightest touch, will be trapped if it causes several gentle movements of a hair-cell. As a rule, there needs to be a second stimulation within half a minute for the two to be associated. The sequence becomes complex:

1. Movement of a hair-cell, if slight, may be ignored; if sufficient, it causes a receptor potential to fire.

2. If this is small, it is ignored; if it is sufficiently strong, an action potential will fire and an electrical charge will pass across the open trap.

3. If the action potential is insufficient, the trap remains open; if it is strong enough, the trap will snap shut.

4. A later stimulation of a hair-cell may be too weak to generate a response. If a further stimulation within half a minute is enough to trigger a potential, the trap will close.

There are several stages of information processing in the Venus fly-trap. The timing of the responses, together with their nature, shows that they are based on mechanisms much like those in animals. Although the fly-trap is well-known, there are several other plants which move just as rapidly. The water-wheel plant *Aldrovanda* lives in ponds and is widely distributed across the globe. It is a pond-weed too small to be studied by the naked eye, and grows in the form of little whorls of radiating leaves. Like the fly-trap, the leaves lie open but can snap shut when a trigger hair is stimulated by passing prey. Because they are so small, you can see a trapped insect inside them through the semi-transparent leaves. The trigger hairs are proportionately longer than those of *Dionaea*, but because the traps are smaller they close much faster.

There is an even more efficient trapping plant, the bladder-wort *Utricularia*. This is a small submerged plant, forming minute bladders which grow out from its leaves. Only the tiny yellow flowers ever reach the surface, for the plant itself stays submerged in the peaty water where it grows. The trapping mechanism is different from that of the species we have considered so far: each of the bladders is closed with a little hinged door, and the cells within the bladder absorb some of the water so that the whole chamber is under negative pressure. Trigger hairs grow from the trap which closes each bladder, and when one of these is stimulated the door to the bladder springs open. That is the opposite of what happens in the fly-traps: they spring shut, while the bladderwort springs open.

The opening of the door means that there is a sudden inrush of water into the bladder, and the little water-flea which touched the trigger is swept inside with the water. The trapped creatures break down and decay, and the soluble components are absorbed by the bladderwort as it grows. Bladderwort is amazingly efficient: the bladders of a small plant can become filled with hundreds of trapped water-fleas in a short period of time. Because these intricate plants are small, they are not as well known as the larger fly-traps. But the production of receptor and action potentials shows that the way these plants operate is close to the mechanisms we see in the animal world. Were plants not confined by their stiff cell walls of cellulose, they might match animals for speed. The study of electrical activity in plants shows that they can sense touch, and respond to the stimulus, in a coordinated and appropriate manner. Clearly, they process data. They do no more than they need to, but what they do is well adapted to the demands of their daily lives.

The tender trap

Some carnivorous plants move more slowly and deliberately. *Drosera*, the sundew, is a common plant in swamps and boglands. There are many different species, some tall, others prostrate; some have long sticky leaves, while others have leaves that are as round as pin-cushions. The trap mechanism involves no movement at all. Instead, the leaves of the sundew are covered with a dense layer of hairs that stick out at right-angles from the leaf surface. The hairs look like dressmakers' pins, for each one is topped by a round, sticky bead which attracts insects and holds them firm.

Once the prey has become lodged on the glue which tips each leaf-hair, the other hairs nearby start to curl over towards it. They help to hold it in place, and also secrete a juice rich in enzymes which digest the tissues and free the nutritious contents for the plant to absorb. Recent research shows that there is a receptor potential created when an insect stimulates one of the sticky hairs. If a hair is stimulated by stroking, the receptor potential causes a whole series of action potentials to cascade down the length of the hair. The stronger the stimulus, the higher the action potential. So, although the movement is slow and deliberate, the mechanisms that cause the bending are complicated. They are clearly related to the strength of the stroking stimulus, and are similar to the way some animals (like sea anemones) respond to touch. Although they do not have nerve fibres, these plants manage to mimic many of the mechanisms of animal sensation. However, the controlling mechanism for the curling over of the leaf hairs remains much of a mystery.

A similar plant is the butterwort, *Pinguicula*. It has oval leaves covered with glandular tissue which secretes a sticky substance that attracts flies. Their presence stimulates the edges of the leaves to curl over, so that the insect is surrounded by digestive glands. Interestingly, in this case it seems that the main stimulus is not the formation of an action potential, but the presence of organic molecules on the leaf surface. These plants are found in peaty soil, or living in peaty water. There is a reason for this. Peat forms in acidic conditions, and the levels of nutriment in the water can be too low to support much plant life. Supplementing the diet with insect protein allows plants to grow in conditions they might otherwise find it hard to endure.

Other insectivorous plants are larger than any of these, though they do not rely on movement to trap or hold their prey. These include the pitcher-plants of the genus *Nepenthes*. The leaf in these plants forms a rounded covering to a flask-like chamber which grows from the flattened petiole (the stem below the leaf itself). The lining of the pitchers is attractive to flies, with lurid splashes of purple and red. Inside the pitcher is a lining of button-like glands which secrete a digestive juice; the juice forms a deep pool at the bottom of the pitcher. As the flies enter the chamber, they are attracted by the liquid and eventually find they are unable to escape. Large numbers form a nutritious broth inside the pitcher. Sometimes the accumulation of half-digested insects is so great that birds peck open the pitchers to feed on the contents. The pitcher-plant is perhaps the most grand and glorious of all insectivorous species, but movement is not part of its carnivorous activities. Clearly, it doesn't need movement. The pitcher plant is doing well enough without.

Mechanical movement

Some of the movements in the plant kingdom result from harnessing the mechanical properties of matter. One of the earliest to appear in the literature of science was the awn of a grain of oats, a flexible spine at the end of each grain which twists and untwists according to the humidity of the air. Robert Hooke (1635–1703), the natural philosopher who is commemorated by Hooke's Law, used a wild oat awn to measure humidity and drew the apparatus in his pioneering book *Micrographia*, first published in 1665. This marked the invention of the hygrometer. Some seeds harness the movement induced by changes in moisture to propel themselves into the ground. The storksbill *Erodium* has spiral awns which coil and uncoil with changes in humidity. They propel the seeds across the surface of the earth until they encounter a crevice, and then the twisting movement helps to screw the seed deep into the earth where it can germinate in seclusion.

The drying out of fern sporangia is used to propel spores away from the plant like stones from a catapult. These sporangia are fringed by a layer of peculiar thickened cells, the annulus. Most textbooks explain its function incorrectly. They tell how the drying of the sporangium causes the cells of the annulus to contract suddenly, breaking open the capsule and scattering the spores far and wide. That is not what happens. In reality, the drying causes the capsule to break as the catapulting collar slowly bends backwards. As water evaporates from the cells they contract ever further, until the whole structure is bent back upon itself. Suddenly, the partial vacuum inside each cell of the annulus is more than can be endured. The cell walls rupture, air rushes in, the shock-wave causes all the other cells to rupture in sequence, and the annulus springs back to its former position with a dramatic jerk. The effect is to launch the spores at high speed, all in the same direction, well away from the fern on which they formed. They travel far enough to reach currents of free air, and this completes their distribution.

Mosses have spore capsules surrounded by teeth which respond to moisture. As the humidity changes, the teeth writhe back and forth and help distribute the spores into air currents. Liverworts have cells which form amongst the spores. These have spiral structures which respond to humidity and twist and untwist, dislodging the masses of spores and freeing them for distribution.

Flowering plants sometimes resort to explosive mechanisms for seed distribution. The most conventional form is a seed-pod which is sprung against itself, so as it dries out it becomes stressed from within, ready to burst at the slightest touch. *Geranium palustre*, the marsh geranium, is one plant which shoots seeds out when the capsules spring apart on being touched. Perhaps the most spectacular is the riverside plant *Impatiens nolitangere* (literally: 'touch-me-not') which has beautiful blooms reminiscent of those of an orchid. The seed capsules are elongated and bright green, maturing to a reddish brown. At this stage, if the capsule is touched it bursts open, the sides curling backwards, shooting the seeds out in all directions at high speed.

Other plants have a more directional 'gun'. An example is the squirting cucumber *Ecballium elaterium*. Pressure builds up as the fruits mature on this plant, and suddenly the entire fruit breaks free. It shoots away, propelled by the pressure of mucilage and seeds squirting from the stalk end, for all the world like a rocket-propelled device. The longest distance one has been seen to travel is 13 m (40 ft)!

Exactly the opposite happens with the dwarf mistletoe *Arceuthobium* which parasitises forest trees in the USA. Most types of mistletoe use sticky seeds for dispersal, which stick to birds and are carried away to a new host. The bird deposits the seeds by scraping its beak on the bark to clean itself after feeding. *Arceuthobium* has a more dramatic method. The fruits are filled with seeds lying in a liquid. The pressure builds up as the seeds mature, until the end of the fruit ruptures and the seeds are squirted out at speeds up to 100 km/h (about 60 mph). They can easily reach the next tree, and this form of parasitism has caused heavy losses to the timber trade.

Many flowers have a similar spring-loaded mechanism. The leguminous flowers (such as peas, beans and alfalfa) hold their stamens between paired petals which form a keel at the base of the flower. If small insects alight, nothing happens. It takes a bee of the right size to bear down on the keel. When that happens, the petals burst open and the stamens shoot upwards like an uncoiled spring, dusting the insect with pollen. This is clearly an unpleasant experience, and it has been shown that many bees have learned how to deal with the plants: they alight on the side of the flower, and take the nectar they require by reaching cautiously between the petals.

Flowers that remain unpollinated by insects (those grown in greenhouses, for example) sometimes spring their anthers without any outside interference. It is as though they have retained a memory of what they are meant to do. There are aspects of plant behaviour that seem to suggest that a plant can store a memory of an earlier experience. Plants have been shown to remember earlier traumas – they can recall being wounded on one side, and compensate for the damage by later growth. If you have dandelions in a mown lawn you may observe that they flower almost prostrate on the ground, as though they have learned that a raised profile will lead to them being cut off in their prime. Equally interesting is the fact that plants can distinguish between one stimulus and another. If a sensitive plant is repeatedly stimulated by touch it will eventually fail to respond. If another form of stimulus is applied (an electrical stimulus, say) it will immediately respond to that. Plants clearly have the means to tell different types of signal apart. More research is needed into these phenomena, which suggest that plants are more alive to their surroundings than is widely believed.

Plants and movement

There are many plants which move parts of themselves much like animals. The common rock-rose *Cistus* is an example. This attractive flowering plant is common near the coast, and grows on headlands buffeted by wind. The rock-rose is pollinated by insects, and clearly could risk losing its pollen to the blustering sea breezes. The flowers have evolved the perfect solution. In normal conditions the stamens are held together in a clump. In this way pollen from the anthers is prevented from being lost to the vagaries of the wind. When an insect lands on the flower, the stamens respond by opening like a rose in a time-lapse movie. Hold the fingers of one hand together in a group, fingertips touching, and then open the hand wide – that's the speed at which the rock-rose opens its anthers and surrenders pollen to a visiting insect. Plants that can show movement are not all exotic, rare or famous. They grow all around us.

Sensitive plants are the best-known examples, one of which, *Biophytum*, folds its leaves as an insect approaches, yet before it actually makes contact with the leaf. It is believed that the sensory cells are responding to static electricity carried on the insect's body, rather than the down-draught from its wings. There is a rambling vine-like plant which grows in the southern USA called *Schrankia*. It is covered with thorns which are ordinarily hard to see, because of the covering of luxurious vegetation. When touched, the leaves collapse, leaving the spines prominent. This is assumed to be a protective mechanism, designed to deter a would-be grazing animal. Lianas, widespread in tropical rainforest, close their leaves when it rains. The stimulus of raindrops on the leaves of one of these climbers, *Machaerium*, makes them close up quickly when it rains and thus gain some protection from the force of the water. The grooved form taken by the leaves when closed helps them to act as gutters, draining the water efficiently from the plant. Some common plants of cooler woodlands – like the wood sorrel *Oxalis* – can also close down their leaves when it rains, though they respond more slowly.

One plant, above all, is well known as the sensitive plant. It closes its leaves in a second, and has long been known for this remarkable phenomenon. If a leaflet is touched, the paired leaves close up like an array of dominos falling in sequence, and the entire plant shuts up rapidly. This is *Mimosa pudica*, though there are several other species which are just as touchy. There are many old stories of the behaviour of this plant. One famous account was written by Robert Brown (1773–1858), the Scottish surgeon-turned-botanist who gave the cell nucleus its name in 1828. He described how a walker passing through a group of these plants might find them closing down in response to footfalls on soft peaty ground. As Brown wrote, even a galloping horse on a nearby path, or a train rattling along the track, can send sufficient stimulus to have the *Mimosa* plants closing up like umbrellas.

There are two remaining problems with *Mimosa pudica*. One of these is where its sensitivity resides. There are no sensory hairs, and no specialised cells that seem to be specifically adapted to detecting pressure or touch. The cells of the leaf seem much the same as cells in any other plant. And, though there are several other sensitive species of *Mimosa*, others show no such movement at all. *Mimosa dealbata* looks very similar, for example, and shows similar cells under the microscope, but lacks any sensitive movement. The second major problem is the purpose that the collapse of the leaves might serve. The rock-rose moves its anthers to protect its pollen; the fly-trap springs its leaves shut to capture prey. Without these mechanisms, the plants would find it hard to survive where they grow. This does not appear to be the case for *Mimosa pudica*. Many theories have been put forward – protection from the sun, avoiding being grazed by passing herbivores, and so forth – but they would apply equally to any other species. If those were the reasons, we would have leafy plants drooping and closing wherever we looked. We don't, of course. These mysteries remain.

The earliest research into the mechanism behind the sensitive plant showed that contact produced a collapse in the cells that ordinarily hold the leaves erect. As water passed rapidly out of these supporting cells, it was believed that they collapsed mechanically inwards and allowed the leaves to droop. A few years after the historical experiments by Burdon-Sanderson, who measured electrical responses in the Venus fly-trap, a German physiologist – Karl Kunkel of Heidelberg – described similar electrical impulses in *Mimosa pudica* as the leaves moved. This failed to trigger an upsurge in research, because it was concluded that the electrical activity was the consequence of the collapsing cells, and not the cause.

In recent years we have begun to find many other clues to the nature of the movement, and some of them point towards a distinctly animal-like series of mechanisms. There are long, thin cells inside the tissues which conduct sap, and some people have likened them to a kind of 'nerve' fibre. Tannin was discovered within the cells, and tannin is a concentrated source of potassium ions, which is important for movement in animals. It has been discovered that, when the supporting cells collapse, there is a sudden surge of potassium through the cell membranes which causes them to lose water and collapse. Tiny vacuoles, fluid-filled spaces within each cell, have been found which have the ability to squirt water out of a cell at speed. More surprisingly, there are known to be fine fibrils inside the cells, which seem to be able to contract like muscle cells in animals.

Why the sensitive species of *Mimosa* move so rapidly, while other very similar-looking species are unmoved by the sense of touch, is still unknown. Meanwhile, it is clear that these curious plants have mechanisms which are parallel to those we see in animals. They have a remarkable sense of touch, and respond much as animals would respond.

Refined senses

The sense of touch is highly influential in the life of many flowering plants. It is at its most highly developed in climbing plants, which have developed an extraordinary sense. Tendrils are highly adapted organs used by many plants to support them as they grow. Most tendrils are a few centimetres in length, but those of the grape vine *Vitis* can measure 50 cm (20 in). Normally, the tendrils move slowly round in an oval pathway as the plant grows up, as though searching for a point of contact. On a speeded-up time-lapse film the effect looks startlingly like a blind creature feeling its way. There is a sense of orientation to these searching movements. For example, the pea plant moves its tendrils in an ellipse so that the long axis of the ellipse is always at right-angles to the sun. If an outstretched tendril comes into contact with a solid support, it will slowly grow to enclose it and thus to support the plant. This response is called contact coiling, and it is easy to speculate how it might work. The side of the tendril leaning against the support is in the dark, let's argue, so that side grows more slowly than the opposite side which is still in bright light. In time, this unequal growth will cause the tendril to grow in a curved path until it has encircled its support and is holding on tight.

That is easy enough to understand, but the phenomenon is actually much more complex. Once the tendril has attached itself firmly to a support, the rest of it keeps coiling up, like the cable to a telephone handset. As it shortens, it lifts the plant closer to the support and actually raises it as it grows. Many plants have a highly developed response to touch. The record-holder is said to be *Cyclanthera pedata*, in which the curling begins within a second of being touched. In some species of the passion flower* *Passiflora* (particularly *P. gracilis* and *P. sicyoides*), the tendrils will be coiling up within half a minute of being touched.

Pea plants are convenient for casual study, for their tendrils can store the memory of a stimulus. If a pea tendril is stroked it will start to curl, though if the plant is chilled this response does not occur. The memory remains, however, and if the plant is later allowed to warm up, the tendril then curls as if recalling or having stored the earlier stimulus. The tip of a pea tendril

---

* The fruit of the passion flower, or alternatively passion fruit juice, is often taken as an aphrodisiac. Wrong passion. The name comes from the passion of Jesus at the crucifixion. These beautiful and striking flowers have twelve petals (the disciples), a lobed pistil rather like the cross, and a ring of pointed purple anthers said to symbolise the crown of thorns. Those who consume the fruit as a sexual stimulant are misleading themselves, I fear.

grows into a sharp little hook, helping to attach it to its support. If you stroke an outstretched pea tendril you will see it start to coil within a minute or so. Try it at night and nothing happens. The tendrils need to be in the light before they will respond to stroking. They can store the effect of the stimulus for more than an hour and, if brought into the light 90 minutes after the stimulus, they will start to coil as though they had just been touched.

The sense of contact has more complex consequences in many climbing plants. Ivy climbs a wall by sensing its direction and growing adventitious roots along the surface it touches on the way. Once the roots touch the wall they hold fast to the surface and grow intimately into the tiny pores and irregularities of the stonework. If you strip ivy from a wall you will usually find that the roots stay behind and remain firmly united with the wall for years afterwards. This response is carried to extremes in plants like the Virginia creeper *Parthenocissus*, where the contact response is to spread the end of the growing rootlet into a disc which becomes a firm anchor to the supporting wall.

If even a single touch-cell is stimulated, the effect is transmitted to all the other cells in the tendril, so coiling starts simultaneously all along its length. These cells can clearly communicate with their neighbours. The sense can be more highly developed than the sense of touch in humans. The touch of a single wisp of wool, less than you can detect on your skin, is enough to start some tendrils responding. The organs of touch in humans can detect a fine hair weighing 0.002 mg drawn across the skin. The sensitive hairs of *Drosera*, the sundew, can detect a stimulus of 0.0008 mg, while *Sicyos* tendrils respond to 0.00025 mg, which is eight times lighter than humans can detect. Not only have plants the ability to sense what's going on, but some do it far better than we can.

The electrical nature of the stimulus has been demonstrated

in several ways. There are action potentials which can be measured in a stimulated tendril, for one thing; and if an electrical signal is actually fed to a tendril, it can itself induce coiling. The pioneering experiments by the Italian physiologist Luigi Galvani (1737–98) at the University of Bologna in the late eighteenth century showed that electricity could stimulate frog muscle and make it twitch. Now we have made similar observations in the tendrils of flowering plants.

There is a further comparison between plant and animal movement, namely that plants can be anaesthetised much like humans. It has been known for many decades that a dose of ether, chloroform or morphine can render a plant senseless. *Mimosa pudica* ceases to respond to touch, and the moving stamens of flowers cease to change position when stimulated. Charles Darwin showed that even a Venus fly-trap can be anaesthetised so that it fails to respond to repeated touching of the sensory hairs. In time, when the effects wear off, these plants return to their normal response pattern. The fact that we can anaesthetise plants, like animals, is a further intriguing point of comparison.

Climbers that twine

Twining plants usually attach themselves to a support through a growing tip which spirals upwards. Once it has encircled a supporting stem, the plant grows taller, encircling the support as it grows. Most of these plants coil in an anticlockwise (left-handed) direction, though a few are clockwise (right-handed). In the hop *Humulus* and the honeysuckle *Lonicera* the twining motion is clockwise. The humorists Michael Flanders and Donald Swan once wrote a light-hearted song (a gentle political satire) about the entanglement of right-handed and left-handed climbers. This is correctly rooted in science, for the bindweed *Polygonum convolvulus* is unique in having separate plants which can coil in either direction. In some species, including the morning glory *Ipomoea jucunda* and the nasturtium *Tropaeolum tricolorum*, coiling can be in either direction on the same plant. This is a reflection of the growth habit of the plant, but is it controlled by the senses? In one twining plant, a parasitic climber, *Cuscuta*, the dodder, we know this to be the case. In this species it is known that the stems respond to touch, for this is the stimulus that causes them to start entwining the host plant.

The rock-rose *Cistus* is not unusual in having spontaneous movements within its flowers. Many other flowers have sensory organs which make them move. Barberry flowers, of the genus *Berberis*, have stamens that move, except that they are ordinarily spread wide apart and clump together when stimulated (the opposite to the behaviour seen in *Cistus*). So does the winter-flowering *Mahonia*, which has holly-shaped leaves, but is a relative of *Berberis* and not of holly. This plant has waxy flowers which appear in midwinter and emit a lingering odour reminiscent of orange-blossom, while the flowers of the barberry are an orange-yellow hue. In both types the stamens are normally spread out around the central pistil. When an insect alights, the stamens close up rapidly and dust the fly with pollen, ready for transportation to another flower.

Studies of moving stamens reveal many strange aspects of behaviour that are mysteries to science. The stamens of the sun plant *Portulaca*, a cultivated garden flower, respond to touch within a few seconds. Unlike the bunched stamens of the rock-rose or the barberry, which simply spread wide like an opening fist, the stamens of the sun plant turn towards the stimulating insect, no matter on which side it lands. Experiments show that if both sides of a stamen are touched at the same time, no movement results. Could this be because each side tries to contract, resulting in the stamen remaining straight? Perhaps; but if the stamens on one side of the flower are stimulated and start to move, the stamens on the opposite side – if touched immediately afterwards – remain perfectly still. Clearly, the action of the bending stamens must be able to communicate their response to the rest of the stamen group.

Stamens that bend towards a fly are surprisingly common in plants, but so are stamens which shorten to expose the female stigma within. The cornflower *Centaurea* has stamens which contract, and so do most asters. They can sense the arrival of an insect and pull back in response. The stigma, which receives pollen for fertilisation, is normally hidden between the heads of the anthers, but after stimulation the anthers draw back so that the head of the stigma is exposed. In this way a visiting insect can deposit pollen while collecting a new supply of pollen for a future visit to a neighbouring cornflower.

Many plants have perfected this response, so that pollen is actually squeezed out of the anthers onto the visiting insect like whipped cream from an aerosol. In the trigger plant *Stylidium* the response is seen at its most dramatic. It has a specially adapted sexual structure in which the male and female organs are fused together into the trigger from which the plant derives its name. The trigger is ordinarily bent away from the centre of the flower. When an insect alights, the flower senses the movement and the trigger is activated. With a sudden movement lasting less than one-tenth of a second, it swoops across the flower and deposits a dash of pollen on the hapless visitor. At the same time, any pollen on the fly is collected by the stigma hidden within the trigger.

Flowers as traps

Some flowers are adapted to use this form of movement to trap insects, though why we cannot say. Dogbane, *Apocynum*, has lively and responsive stamens which trap insects. As a fly alights, the stamens trap it and prevent it from escaping. The fly struggles for release, but eventually dies of exhaustion. Perhaps this mechanism is destined to evolve towards a way of obtaining supplemental nutrients from the body of the dead fly, or perhaps it evolved originally as a way of trapping flies so that pollen is borne from one flower to another. The dogbane plant becomes discoloured by the dead insects each flower has ensnared. At present, there is no known explanation for this behaviour.

In other cases, flies are trapped for a purpose. The cuckoo-pint *Arum* traps insects in a manner that has much in common with the way pitcher-plants trap insects for food. These plants have an attractive spathe, a bright red rod in the centre of the curved flowering body. From the spathe arises an odour of rotting flesh, which some flies find attractive. The flies are attracted by the sight and smell of it, and are lured down into the flower, past a series of backward-pointing hairs, towards a nectary hidden in the base. Those flies already bearing pollen thus fertilise the stigmas hidden inside the flask-like base of the flower. They find they cannot get out, and are kept alive by the nectar within the flower. They are allowed to escape only when the hairs have withered and the insects can emerge, covered with fresh pollen for the next *Arum* in line.

There are many other flowering plants in this family, of which the most dramatic is the voodoo lily, *Sauromatum guttatum*. It stands tall, its lurid speckled hood visible for a great distance. The voodoo lily emits the odour of rotting meat, the perfect lure for flies, and even heats up so that it perfectly imitates decomposing flesh. When it is approaching maturity, the flower becomes warm to the touch (some 15°C [27°F] hotter than the rest of the plant) and burns energy at an astonishing rate. It is certainly comparable with the metabolic rates of a fast-moving mammal. The hungry flies go straight to the centre of the flower where they are held by the halo of backward-pointing hairs and are released only when the pollination cycle is ready to proceed. So much energy is released that the plant's metabolic systems are stretched to the utmost. The hormonal control is elaborate and, as we shall see, it parallels what we find in the animal world.

Sexual movements occur in many plants. The familiar garden plant *Mimulus*, the monkey flower, has a bloom in which the female stigma is divided at the end into two open lips. If either is touched, the two close together, enfolding any pollen that might have been deposited. If pollen is present, and the flower is fertilised, the lips stay closed; but if fertilisation does not occur, then after a while the two lips part again, waiting for the next visit by a pollen-bearing insect (or, in this case, a curious human observer). Movable female sex organs are found in many other flowers. They occur in the butterwort and the bladder-wort, two of the carnivorous plants we considered earlier (see pp. 194 and 196). The cultivated garden plant *Incarvillea delavayi* has stigmas which close when touched, and in these an action potential has been found to occur in the split-second before the movement begins. There is even a slight movement in the stigmas of the foxglove *Digitalis*. The fastest closure is seen in *Martynia*, which takes only two or three seconds to close its stigma, while it can take up to ten seconds for closure in the monkey-flower.

Pollination movement is most highly developed in some members of the orchid family. The orchids have beautiful blooms of a highly specialised nature. Their petals are scooped and fused into an attractive, hooded structure. The opening of the flower and the surrounding petals is said to be reminiscent of the pudenda of the human female, and there is an enduring mystique about orchids the world over. Orchids have the small-est of seeds. They are like fine pepper-dust in appearance, and are as small as many pollen grains. Some orchids have a highly developed sense of touch. They have a specially adapted sensitive petal, the labellum. As a fly alights, the stimulus sets the labellum into action. It rapidly folds down, trapping the insect inside. There is an opening visible to the captive fly, but as it struggles free it becomes coated with pollen, while any pollen already present is picked up by the orchid's stigma. These phenomena were first studied by Charles Darwin at his Kent home, Downe House. In a nearby field many species of orchids grow and this started him on an investigation into the numerous ways in which orchids use insects for cross-pollination.

Many orchids mimic insects. The bee orchid *Ophrys*, an annual, is one example. It attracts insects of the same species as the one it imitates. One of the Australian orchids, *Drakaea*, has a specially adapted labellum which not only looks exactly like the female of a species of wasp, but emits the same odour as the wasp, as a lure. Males searching for a partner are attracted to the orchid, where they try to mate with the camouflaged labellum. No sooner do they take hold than the specialised petal springs into action and throws the insect deep into the bloom. Any adhering pollen is collected by the orchid's stigma, and two of the sticky masses of pollen from its own anthers are pressed against the wasp's body. As it travels from flower to flower, searching for a mate, the male wasp may collect several consignments of the sticky pollen. These pollen masses, known as pollinia, are highly adapted in many species of orchid. Not only do orchids have curious flowers and strange pollination mechanisms, but they too have a well-developed sense of touch.

Forecasting the weather

Tales about flowering plants that move with the weather go back to ancient times. The scarlet pimpernel, *Anagalis arvensis*, has been said for centuries to act as a weather forecaster, closing its petals if rain is to follow. Observation of this little plant, with its bright red flowers, shows that it always closes its flowers in the early afternoon, whatever the weather. Perhaps it was the occasional afternoon rains of a hot British summer which gave rise to the association in folklore.

Wood sorrel, *Oxalis acetosella*, has delicate lobed leaves like those of clover. They droop and close as evening approaches and are shut at night. It is also said that they foretell the rain by closing up before rainfall begins. The movement of the leaves of wood sorrel has the appearance of a protective response. *Oxalis* leaves can be induced to close by the sense of touch, though it takes a good deal of stroking (and many minutes) before anything happens. The leaves seem to be sensitive to temperature levels and certainly they close when heavy drops of rainfall knock the leaves and stir them into action. Perhaps this is how the tales of the forecasting ability of the plant first arose. As we shall see, the leaves of wood sorrel are also sensitive to light – too much, and too little. Many other plants have this apparent ability to 'go to sleep' at night. The plants in a garden, a stretch of woodland or a meadow often adopt a very different configuration in the dark from how they appear in sunlight. The closing of flowers and the drooping of leaves in the dark have been recorded since the most ancient accounts of plants were written. The tamarind tree, *Tamarindus indica*, folds its leaves at night, as does the telegraph plant *Desmodium gyrans*. In sunlight the leaves of the telegraph plant stick out at right-angles to the stem, but in the hours of darkness they hang straight down. *Desmodium* also has small, leaf-like projections which jut out from the stems. These stipules move round in circles as though sending messages. The rate of rotation increases with the temperature, and on a hot afternoon a single stipule can rotate in a circle in a minute or two. The movement is clearly visible to a patient observer who is prepared to sit and watch. Early telegraph machines had magnetic pointers which swung up and down as messages were tapped out and the plant is named after these machines. The movement is elaborate and regular, yet nobody has thought of a reason why it might occur.

Simple folding movements in response to a general stimulus (like light levels) rather than a specific stimulus (like touch) are known as nastic movements, or nasties. The behaviour of the leaves of the telegraph plant is an example of, in scientific terms, a *nasty*. It's an odd term, not a joke word, and means 'folding'. Plants do not respond to light in the same way; we have seen that *Oxalis* opens its flowers during the day and closes them at dusk, and the scarlet pimpernel opens them at dawn and closes them after lunch-time. Other species avoid the sunlight; the evening primrose, for instance, keeps its flowers closed during the day-time but opens them as dusk approaches. Because these movements are a response to the plant seeing how bright the light is, they are classified as photonasties.

Thermonasties are the movements in plants caused by a change of temperature. The familiar rhododendron evolved in high mountains where low temperatures are common, and evolved a nasty to deal with the cold. Normally the plant has outstretched leaves like those of a laurel, but when the temperature falls below freezing they curl inwards and roll up; each leaf also droops towards the ground. In this way they can reduce the chance of frost damage. Not only do *Oxalis* leaves close when rain falls on them, but they also respond to nightfall and remain closed during the hours of darkness. Interestingly, they can also monitor when it is too bright. In direct sunlight the plants respond by closing their leaves, just as though it were raining. In this way they can see when it's too bright, and can take action to prevent damage to the leaves through overheating.

Although plants like light, and green plants need it to grow, too much light is often just as great a hazard. One of the simplest but most dramatic examples is the compass plant *Silphium* which grows on the US prairies. As flat new leaves grow, they detect the direction of sunlight and develop in a north–south alignment. All the leaves of every plant are parallel to each other. This means that as the sun rise, it shines directly onto one side of a leaf, giving it the full benefit of the solar energy. As the sun rises towards noon, it moves round so that it is shining edge-on to the leaves, which are thus protected from the full heat of the midday sun. During the afternoon, the sun moves round so that it starts to shine on the opposite side of the leaves, and they receive a second intake of solar energy. The leaves do not move, but their careful orientation means that they can extract the sun's energy throughout the morning and afternoon while avoiding the risk of over-heating during the heat of noon. The compass plant manages to have a siesta through its curious design.

A sense of electricity

Plants have another sense of which we have little knowledge – the ability to detect an electrical field. There has long been a belief that a lawn becomes greener before the torrential downpour of a thunderstorm. This belief is not new. Early in the twentieth century, experiments were carried out where high-tension cables were stretched across a field of growing crops. The results showed that the plants did indeed 'green up' in response to an electrical field. During the 1990s, research at Imperial College, London, located sensory cells within the plant which have the ability to sense electricity. It is true – plants really do 'green up' in thundery weather before the rain starts.

The physiological purpose of this is complex. Before a dried-up plant can fully benefit from rainfall, a great cascade of enzymes needs to be set in train. The dried leaves need to return to an active state of metabolism, ready to receive the water when it comes. This takes time, and when rainfall is due, the sooner the process starts the better. Plants have adapted to this fact. When an electrical storm is approaching, cells within the grass leaves begin to mobilise their metabolic processes, ready for the rainfall. The lawn really does turn green and, through the grass's extraordinary senses, it does so *before* the first drops of rain begin to fall.

Solar input

Avoiding the sun may be important for some desert species, but plants in cooler latitudes need to collect sunlight whenever they can. In these species we find a regular daily movement as the plants track the sun across the sky, turning to keep facing it from dawn to dusk. Some plants are named for this legendary ability, like the sunflower and the heliotrope (*helios* is the Greek for 'sun'). The leaves of cotton plants follow the sun, and *Malvastrum*, a plant of the deserts of California, also has leaves which it keeps facing the sun from dawn to dusk. These are tough species well used to the rigours of a hot climate. Lupins like *Lupinus arizonicus* follow the sun too, though they have more delicate leaves and manage to avoid over-exposure by turning their leaves aside during the hours around noon.

Northern plants need sunlight more than any others, for it is in short supply and the flowers of the Arctic tundra need all the light they can get. One garden plant which is found growing wild in northern and western Europe, *Dryas octopetala*, shows how important it is to follow the sun. An experiment showed what happened if you fix the flowers to stop them turning through the day. In the plants whose flowers were free to follow the sun, the internal temperature rose 1°C (1.8°F) higher than in those whose flowers were prevented from moving. This is an important difference, for it turns out that the size (and hence the viability) of the seeds is a function of the internal temperature of the flower. Thus, the ability to turn and face the sun can affect the chances of survival for plants.

In a few species, the behaviour depends on the conditions under which the plant lives. An Australian plant grown as a forage crop called siratro, *Macroptilium atropurpureum*, will orient its leaves so that they face the sun when the ground is rich in water. During conditions of drought, however, it changes its behaviour and holds the leaves edge-on to the light (like the compass plant). Clearly, this is meant to reduce the amount of evaporation from the leaves when ground-water is in relatively short supply. These mechanisms are present in a large number of plants. Even those which do not actually turn to follow the sun are able to sense the direction and strength of light, so that new leaves do not overshadow older ones more than necessary. The mosaic of leaves in trees, or on plants like ivy or Virginia creeper, is carefully contrived to give each leaf a fair share of the light. In that sense, all plants can see and can grow to optimise their benefits from sunlight. Before concluding that this is a simple physical mechanism, like a magnet attracting a pin, we need to understand that this turning mechanism sometimes suggests that plants have some form of active memory.

Plants become even more interesting when we consider not what they do during sunlight, but how they behave in the dark. Many of the species that follow the sun prepare themselves each night for the dawn. They turn to face the direction where the sun is due to appear. *Malvastrum* provides an intriguing example. The leaves are kept facing the heat of the sun all day long, finishing up at sunset facing west. After the sun goes down they resume a normal posture with the leaves spread out conventionally and facing upwards. As dawn approaches the leaves turn again to face the east, ready for the time when the sun will appear. It is not unusual for sun-following plants to turn towards the direction of dawn during the hours of darkness. An interesting experiment would be to grow some plants in pots, and turn them through 180°. That would show how quickly they could learn where the sun was due to rise, and it would demonstrate whether they were truly remembering the direction, or simply sensing a change in the light.

How plants see

The sense of light detection exists even in seeds. Many plant seeds require sunlight to germinate. This explains the poppy fields of the battlegrounds of the First World War. Within weeks of terrifying ground attacks, the churned fields were cloaked in blood-red poppies. There was a poetic symbolism to the phenomenon, and it is commemorated in the sale of poppies in aid of British war charities. The disturbance of the soil brought to the surface many seeds which were stimulated to germination. It was the sunlight which made the poppies grow.

An old farming tale suggests that fields ploughed at night grow fewer weeds. Some recent research suggests that levels of weed growth can be cut by 80 per cent by ploughing at night. This does not make immediate sense to me – even if the seeds were exposed at night, they would surely be stimulated by sunlight the next day. However, the facts stand and more research will doubtless unearth the real answer to plants and their reactions to light.

Light is the green plant's primary source of energy, so it is natural that a plant should be able to detect light well enough to ensure that it derives maximum benefit from the sun. Some of the mechanisms of sense are simple. Plant growth hormones in stems are more concentrated in dark tissues than in those exposed to light. The effect of light on growth hormone seems to 'drive it away'. The plant growth hormones, auxins, stimulate cells to grow. Imagine an upward-growing shoot, a simple rod of plant tissue. If light shines onto the left of the shoot, then this lighter side will continue to grow rather slower than before. The increase in the size of the tissues on the darker (right) side will therefore tend to bend the shoot towards the light. Once illumination is evenly detected on all sides, the growth will continue in a straight line. In this way, a shoot will always grow towards the light. If a plant is growing in conditions of semi-darkness the effect of auxins will be maximised, resulting in a fast-growing, elongated, spindly pale plant.

You can observe this effect if a plant germinates down a drain, or inside the mouth of a cave. Seed potatoes sprouting in a garage or a garden shed will develop long, pale, weak stems as they strive to reach the light. This phenomenon is known as etiolation, and though it produces a weakened plant, it assists rapid growth towards a light source. From there on, growth proceeds at more normal rates. The inhibition of growth by light is counter-intuitive: you might expect light always to stimulate growth. However, this is the secret behind the drive all vascular plants show towards a light source and explains why specimen plants growing in dense woodland are always larger than similar examples growing in the open.

There are several related factors which influence the way plants relate to the light which bathes them as they grow. For example, the rate at which the dark side of a shoot develops is faster when light is shining on one side than it is in a shoot grown in complete darkness. Not only can light affect the distribution of auxin within a stem, it can also change the sensitivity to auxin of the cells. Etiolated shoots do not respond to light as rapidly as shoots that are green. If plants are grown in a closed chamber lit from one side, an etiolated plant keeps growing upwards away from the downward pull of gravity. A shoot that has become a little green responds more rapidly to the light, and turns across the pull of gravity to reach towards the lamp. As the pale and etiolated shoot is bathed in the light from the lamp, it begins to acquire a little colour, and once this has appeared it also loses its etiolated appearance and starts to grow towards the light. Plants can sense the direction of gravity and grow up away from its pull, but the attraction of the light overrules this fundamental mechanism in the green shoot of a plant.

Another simple experiment can be demonstrated by covering up a cotyledon – the seed-leaf from a germinating plant – with aluminium foil. If one of the paired leaves is completely covered with foil one morning, so that it cannot receive sunlight, then by the end of the day you can see how the entire plant is turning sideways, away from the direction of the shaded cotyledon. If you do this with several little plants, covering the left-hand cotyledon in each case, you will see them all bending the same way. Even better, cover the left-hand leaf of some plants, and the opposite leaves of others, for then you can see that they bend in either direction, depending on which cotyledon you covered with foil.

The zone at the end of a growing shoot which senses light is not at the very tip, but lies one-tenth of a millimetre (four-thousandths of an inch) behind it. Further down, the sensitivity to light drops off rapidly. In the oat seedling, for example, the sensitivity drops by a factor of 36,000 times over the next 2 mm ($\frac{1}{16}$ in) from the tip. Stems and leaves are generally sensitive to light, but they also respond to heat, and it can be difficult to separate the two effects in experiments. Investigations are hampered by many factors. For example, there are other photo-receptors in plants and more than one of them may be concerned with the response to light. These pigments may interfere with each other and mask a true measurement of their effects. More than one pigment is involved at one time, so that we cannot be sure we have separated out the contribution of the different components, and the light itself varies in quality. Some effects are produced by blue light, and are truly a response to illumination, while others occur at the red end of the spectrum beyond which infrared heat detection also comes into play.

We cannot always be sure what we are measuring. The turning of a leaf or the bending of a stem may be the last in a long line of effects which trigger one another in sequence, so we need to know that a receptor potential can be recognised. In many cases that remains to be done, and the measurements we are making may be a long way down the chain of causality from what originally happened. The plant itself modifies the light in several important respects. Light is attenuated, or weakened, as it passes through the plant tissues. Each photosynthetic pigment captures energy by absorbing light of specific wavelengths, so the light that gets through has a different quality. Light is also refracted, rather as happens with a lens, as it shines through a plant's watery and translucent tissues. Indeed, some plants have developed lens-like structures to intensify this effect. Some plants have a lens-like structure in one end of the cell, which focuses light on a chloroplast at the other end.

The fundamental photosynthetic pigment in plants is chlorophyll, for this changes sunlight into a form of energy plants can use. Through chlorophyll, plants take simple molecules like water and carbon dioxide and fit them together to make carbohydrates. It is a reaction of great value, and is easy to comprehend, yet science cannot imitate it. The solar energy that bombards our planet ceaselessly is more than enough for our needs. With our crude technology, the best we can do is grow a green plant (like oil-seed rape), harvest the seeds, crush them and extract the oil, then burn this oil to provide heat. The heat is then used to drive a car, or power a generator, with further energy losses in the process. It is a chain of inefficient energy conversions which plants put to shame.

The maximum theoretical efficiency for capturing carbon

dioxide through photosynthesis is 34 per cent, and green plants can achieve energy conversion rates up to 90 per cent of this maximum figure. That's only as far as the chemistry is concerned, of course; in practice the amount of solar energy that is captured can never reach this figure. The rate of efficiency at which green plants can collect solar energy and make chemical end-products from it can exceed 5 per cent. There is no harmful wastage, and the only by-product is oxygen. The other photosynthetic pigment in plants is carotene. Carotene was identified near the growing tip of shoots, in exactly the region of maximum sensitivity to light. It has been shown to be vital for the growth of cereal seedlings.

We should note the parallels between these photochemicals and life processes in animals. Where have we encountered carotene before? Yes, in carrots. It is the compound which gives the carrot its distinctive orange colour, and it also provides the bright yellow in the buttercup. It used to be said that you should eat carrots to see in the dark. That's not quite true, but a lack of carotene in the diet causes a predisposition to night blindness. We rely on the photo-receptor rhodopsin to see in dim light, and without the carotenoids in our diet we would not be able to synthesise rhodopsin. Carotene is important for the sense of sight throughout the animal kingdom, just as it is important for plants.

The myth of trees and oxygen

It is widely said that plants (trees in particular) produce the oxygen we breathe, but this is not strictly true. On a bright sunny day, all plants release oxygen into the air. However, their life processes during the dark consume oxygen and release carbon dioxide, much as we do. At the end of its life, a plant rots away. By the time it has completely decomposed, the decomposition process will have absorbed all the oxygen the plant produced during its life. Thus, the life cycle of a green plant does not itself increase the oxygen in the air. This is how the cycle of life operates: the growing tree takes in sunlight, air and water (together with chemical raw materials dissolved in the soil) and creates its tissues from these raw ingredients. As the dead tree decays, its tissues break down again to the original water and carbon dioxide, the chemicals return to the soil whence they came, and the same amount of oxygen given out during its life is reabsorbed by the metabolism of the decay process. The oxygen remaining in our air is what is left over from the green plants and their residues which have not yet decayed: oil fields, coal measures, organic matter in the seas and soil, peat and all living vegetation. Green plants have the crucial ability to capture their own energy from sunlight, whereas animals have to consume ready-made molecules and break them down. This is why all animals consume plants, or animals which in turn have fed on plants. It is the plant that captures the solar energy to sustain life on our planet. Green plants are the single key to life on earth.

Plants and vision

There are light-sensitive chemicals in plants which do not take part in photosynthesis, but give the plant its sense of vision. Phytochrome, for instance, can sense the relationship between red and far-red light, enabling plants to sense the presence of their green neighbours. Another is riboflavin, which was proposed as a light receptor as a result of some simple experiments. If plant tissues are treated with potassium iodide, riboflavin (but not carotene) is inactivated. In that state, the plant fails to respond to light, which strongly supports the idea that riboflavin acts as a sensory compound. Another example is phytochrome. Although there has been much research into this potentially revealing area, it is hard to tell the relationship between a light-sensitive pigment and the way a plant responds. This is where molecular biology could hold the key to unravelling the mechanisms hidden in the green plant's sense of sight.

Riboflavin may have connections with human life. It was originally known as vitamin B_2 and is essential for health. A lack of riboflavin in the human diet leads to problems with the eye. Cataracts can develop, the eye can become reddened and a sense of itching or soreness develops. People deficient in riboflavin find bright light unbearable, and this photophobia can be so severe that they cannot endure even normal daylight. Such interesting coincidences between human sight and the photo-receptors of the plant kingdom remind us of the universality of the senses. The notion that perception is confined to humans, or even to the animal kingdom, can no longer be sustained.

When senescence is new life

In autumn, deciduous trees shed their leaves in a blaze of red, yellow and russet. The phenomenon is seen as a way of ridding the tree of its unwanted, dying leaves. We are taught that the reasons behind the phenomenon of leaf-fall are to help the tree survive winter. The main functions of the fall are said to be:

o The removal of dying leaves.
o Shedding diseased, damaged or old leaves.
o The removal of superfluous foliage from trees that are over-burdened.
o Defoliation of deciduous trees.
o Recycling of mineral nutrients.
o Protection of the tree, for instance by formation of leaf scars.
o Vegetative propagation (the growth of new plants from shed leaves).
o Facilitation of seed dispersal by animals.
o Inhibition of seed germination through substances leaching from shed leaves.

I believe that these standard reasons can be challenged. Diseased and damaged leaves are often shed before the fall. Leaves are shed from healthy trees as often as from those that might be stressed or over-burdened. There is no obvious reason why trees must drop their leaves. Problems with winter weather are no explanation at all, for two reasons. First, evergreens also shed their leaves. We are told that deciduous trees have vulnerable leaves, whereas evergreens have leaves that are specially adapted and so can withstand the winter. But evergreens also shed their leaves. Stand under an evergreen tree and you are up to your ankles in them. Holly may be evergreen, but it sheds leaves in a mini-autumn of its own each spring. If evergreens are physiologically able to withstand the rigours of winter, yet still shed their leaves, there must be deeper mechanisms at work. Second, similar plants show different behaviour. In many plant types there are evergreen and deciduous species with little to choose between them. Some species of the honeysuckle, *Lonicera*, are evergreen while others are deciduous. There is clearly no overwhelming reason that impels a plant to shed its leaves at one time.

We have seen in this chapter that plants are far more complex in their behaviour than most people might recognise. The same is true of this phenomenon of abscission (leaf-fall). The abscising leaf is always described as 'senescent', or ageing; indeed the yellowing leaves of autumn are designated in the scientific literature as 'senescent leaves'. Let us now look inside, to see what the plant makes of it. The key indicator of senescence is approaching death: the metabolic rates slow down until they cease altogether. This is, most emphatically, *not* what happens to the yellowing leaves. Their metabolic rates increase. Activity in the cells changes, but its rate does not diminish. Many of the essential components (such as chlorophyll) are broken down and translocated elsewhere in the plant. These are raw materials being put into store. Meanwhile, other compounds – many of them brightly coloured – are synthesised in the leaf. Waste materials from the plant are systematically passed out into the leaves, ready for shedding. Levels of heavy metals and other toxic materials in the leaves increase dramatically. In some cases, poisonous metals increase a thousand-fold before the leaves are shed.

This is no senescence of the leaves, but a coordinated metabolic process. We all know and recognise that a plant uses its leaves to gather sunlight, and to evaporate water. Here is a further mechanism. The plant uses its leaves to excrete waste products to the outside world. In my lectures and papers on this subject I proposed a new term: that each leaf acts as an *excretophore* (an excretion carrier) as well as a transducer of energy. The autumnal trees which we always dismissed as simply dying down for winter are actually in the throes of new, hidden activity. Those bright hues are the result not of the death of the vegetation, but of the plant actively using its leaves to package its wastes and dispose of them. The compounds in the leaf may help to signal to seeds that they should not germinate, which would produce an overcrowded tree population. Compounds from the shed leaves are used as a metabolic raw material by the choreographed communities of bacteria and fungi in the soil, which offer them, recycled into nutriment, for the next season of growth.

This theory is already beginning to revolutionise the way we look at plants. First there is the theoretical aspect. The concept explains why vascular plants shed their leaves during their lives. It is nothing to do with drought or food shortage; even aquatic plants, with a permanent supply of trace elements, minerals and water, shed their leaves. A palm tree (which would have far more surfaces to capture solar energy if it only retained its leaves) sheds whorls of leaves as it grows, until it is topped by a tiny tuft of foliage. The reason for this is suddenly clear. I believe that this also explains flowering and similar phenomena. Petals and sepals are shed during flowering, and many petals contain coloured compounds. I propose that these discarded organs contain the wastes from the metabolic burden of reproduction, and perhaps some of the compounds within them are metabolic sinks for waste products. Wherever we look in the plant world we see organs discarded after a peak of activity, so I believe that we may be witnessing a universal principle of plant metabolism.

The shed leaves (and other organs) become a crucially important part of the soil. It is true that shed leaves tend to discourage seed germination, but that is often because of the waste they shed which leaches out in the rainfall. Although inorganic chemical residues are returned to the soil by a fallen leaf, so too are the nitrogenous compounds in the leaf, which are then recycled by soil microorganisms ready for later use. If we harness this mechanism we have a way of decontaminating soil through phytoremediation. Metal-tolerant plants, grown in polluted ground, will pass those metals selectively to their leaves prior to abscission. The leaf litter, if collected, would contain the contaminating metals absorbed during the previous year. In time, the metals would be lifted from the ground, harvested by the plant and deposited in leaves, from where they could be recovered by human intervention. I now believe that we could phytoremediate soil in this way, and provide a new source of valuable metals for recycling. Experiments carried out in France and the USA already show that it would work in practice.

When autumn comes, do not look too patronisingly on the beautiful autumn colours. The trees sense the changing seasons and methodically prepare for winter. Their wastes are sorted and processed; harmful metals identified and transported to the leaves. A burst of metabolic activity gives the leaves their new function with a high level of efficiency, enabling them to pass back to the world the specific waste products the tree chooses to discard. If we are wise, we will harness these proclivities to help us restore areas of the globe that are polluted by the activities of human industry. Until then, let us look with respect and understanding at our plant neighbours. They are cleverer than we think.

Plants can sense changes in the weather, and anticipate the arrival of winter. Can they predict the future? Much folklore surrounds the relationship between plant life and the weather that lies ahead. Oak (*Quercus*) and ash (*Fraxinus*) come into leaf at about the same time in the English countryside, for example, and are widely believed to foretell the rainfall for the summer ahead. 'If it's the oak before the ash, we shall have but a splash,' says the first line, while the second warns: 'If it's the ash before the oak, then we shall have a soak.'

Other legends concern the way fruit or flowers form on trees and shrubs, and the way birds nest in them. In my view, the notion that nature is predicting the future in some metaphysical sense is ridiculous, for plants and animals base their behaviour on their past experiences. However, this is not to dismiss the idea out of hand. The science of meteorology is devoted to basing a prediction of future weather on the history of the recent past. In modern weather forecasting, computer modelling of data from sophisticated sensing systems aboard earth satellites and meteorological stations indicates how the weather might behave over the next week or so. Plants and animals may not be able to sense the future, but they can certainly display sophisticated responses to the past. The principle that recent history determines the future is the cornerstone of meteorology, and we can be sure that we are not the only species to detect these mechanisms. If a late, hot summer is statistically related to a short, cold winter, then plants may well produce fruit that are adapted to this eventuality. In this way, nature might be sending us signals about the likely future.

Many plants are sensitive to the length of the day. By exploiting this perception, and growing them in greenhouses under artificial light, we can induce plants to flower whenever we wish. Plants are sensitive to changes in temperature, and we can utilise this response as well. By chilling seeds, bulbs or other parts of plants, we can induce them to develop at unexpected times of the year. This is how we prepare bulbs to flower in the house when winter is still upon us. The senses of plants indicate when it is time to prepare for the winter. The process of preparation is not a mere sequence of senescence, but a complex process of preparation and prudent storage which helps preserve our environment.

Hormones in plants and animals

If photo-receptors have so much in common between the worlds of animals and plants, we should presume that other parallels might exist. It is well known that aspirin is a drug of immense value to human communities. It reduces fever, controls inflammation, helps reduce pain and regulates blood clotting. This quartet of invaluable properties means that, if aspirin were a new discovery, it would be hailed as the therapeutic miracle of the century. Yet it is not new – it is ancient. Aspirin is the trade name for acetylsalicylic acid, and this is named from *Salix*, the willow. There is little mystery about the fact that aspirin originated in extracts of the willow, for it shares its origins with many other herbal remedies. What we rarely consider is this: what was the aspirin doing in the plant in the first place? It was not produced merely for the benefit of human society. If it is present in plants, it must be there for a purpose.

It is becoming clear that aspirin is a plant hormone. This familiar home remedy, for instance, is the hormone that regulates the temperature of the voodoo lily when it heats up as it matures. More remarkably, levels of aspirin have been found to increase fivefold immediately after an attack on a tobacco plant by the tobacco mosaic virus, TMV. The virus takes its time before clinical manifestations appear – the characteristic speckled pattern of pale patches on the leaves – but the rise in the aspirin levels is an immediate indicator that the disease is present. This happens in other plants as well. When cucumbers are threatened by a fungus infection, levels of aspirin increase and stimulate the plant to resist the attack. It is also believed that raised levels of aspirin help plants resist environmental stresses that would otherwise threaten their survival. In animals, aspirin interacts with prostaglandins in the body. Prostaglandins have now been found in a host of different plants, where it is becoming clear that they also play an important role in maintaining metabolic cycles and preserving the plant's health.

Many drugs are extracted from plants and used in veterinary or human medicine. Steroids were first made from extractions of yams. The first birth-control pills were made from these extracts, yet few people realise that the yams themselves had been eaten by native populations in South America since unrecorded times. The women in those tribes knew perfectly well that they could prevent conception; it took Western science thousands of years to catch up. Some of the natural hormones produced by plants affect other animals. Clover contains a substance which mimics the sex hormone oestrogen, and acts as a herbal contraceptive in the wild.

5-Hydroxytryptamine is a hormone in humans which is better known by its more popular name, serotonin. It is a recent fad health-food supplement. Its structure is almost identical to that of indoleacetic acid, one of the auxins (a plant growth hormone). Much of the commercially available auxin is made from serotonin extracted from animal urine, because so much of this 5-hydroxytryptamine is secreted through the kidneys. As we have already seen, abscisic acid (AbA) is a key hormone in flowering plants. Not only does it help regulate the breathing pores, or stomata, but it helps plants manage metabolic stress, and it is a key factor in the shedding of leaves during the fall (abscission being the scientific term for this phenomenon, hence the name). Now, the identical molecule has been discovered in mammalian brains. We do not know what AbA is doing in the brains of rats and pigs, but if it is extracted and applied to plants, they respond as though it was the home-grown version.

A similar pattern can be established with insulin. We know that insulin is produced by the tiny islets of Langerhans cells in the human pancreas, named after the German physician Paul Langerhans (1847–88) who first described them in 1869. Different forms of insulin have since been found to be present in small amounts in many other living organisms, even in bacteria. It is produced by some brain cells. A version of the insulin molecule has now been found in plants. Insulin has been shown to interact with plant cells, and some experiments have suggested that it can affect plant growth.

As I show below, touching or stroking growing plants has the effect of stunting their growth. It is as though they need to grow shorter and more stockily if they are to resist buffeting winds. It does not matter what form the stimulation takes, for they will tend to react in much the same way even if they are sprayed with water. Experiments with one plant, *Arabidopsis*, revealed that the spraying with water activated genes within the plant cells, and the result was an increase in the production of calmodulin. This is a plant hormone with many important functions. Although not widely known, calmodulin can take many forms and has different functions in each guise. It can control the work of enzymes, and regulate gates in the cell membrane. It gets its name from the fact that it modulates calcium ('calmodulin') and it can absorb large numbers of calcium ions. The hormone is also found in animal cells, and calmodulin receptors exist in our own brains. Here we must face an extraordinary truth: not only are similar compounds shared by plant and animal organisms, but in some cases the molecules are absolutely identical. In their current applications they now perform different functions, but their common origins in plants *and* animals are clear. We are far closer to plants than we might think.

Old wives' tales and scientific truth

Many of the discoveries about how plants work can help to explain how some age-old remedies actually work. It has long been said that an aspirin in a vase of water can help cut flowers stay fresh. Aspirin helps motivate the defence mechanisms of plants against infection, so there is one obvious benefit. Aspirin can also help to conserve water (by closing down the stomata through which a plant transpires water) and can stimulate the growth of leaves. It is a perfect drug for treating plants, as well as humans. What about the leaf of the dock plant *Rumex* which is traditionally used to take away the sting of a nettle, *Urtica*? The stinging hairs of a nettle contain formic acid, but this is not what causes the long-lasting tingling and burning sensation. The substances which so irritate the skin include histamine, acetylcholine (a nerve transmitter in animals) and serotonin, which is involved in nervous transmission. Dock leaves contain compounds which inhibit the action of the serotonin, which helps explain the sense behind the legend. Rubbing a nettle-sting with a dock-leaf now has scientific reasoning behind it.

There is a belief that a walnut tree will fruit better if the trunk is struck with a stick. As an old saying put it: 'A dog, a wife, and a walnut tree; the more you beat them, the better they be.' Although the first two are barbarous nonsense, I think there might be truth in the assertion that a walnut tree could fruit more abundantly if its trunk was repeatedly struck. I have put forward the hypothesis that the effect of shock-waves passing through the trunk might lead to the breakage of the fine capillary threads of water that rise through it. Water could not be raised by suction any higher than about 10 m (32 ft), for that is the height of a column of water supported by atmospheric pressure. Trees taller than this rely on cohesion in the vessels which conduct water. The water thread has never been broken, and thus is drawn continually upwards by evaporation from the leaves (and some pressure from the roots), no matter how tall the tree grows. Beating the trunk could cause some of those columns to snap, thus threatening the tree with a loss of water intake. Abundant fruiting is a plant's best defence against such an assault, and it is in this relationship that the origin of the legend might lie.

The walnut tree may yet have other mechanisms by which it could register the assault. Hormones or other signalling compounds could be released in response to the beating, and they in turn could perhaps stimulate the tree to produce additional fruit. I observe that plants often fruit most abundantly when threatened: perhaps by root pruning, by an inclement period of weather, possibly also by beating. If plants were mere mechanical devices, then an agency which harmed or threatened a plant might be predicted to reduce the level of fruiting. In fact, plants have complex mechanisms to react to stimuli and take appropriate action. I find it intriguing that plants can respond to threats by increasing, rather than decreasing, their output of seeds.

What of the people who talk to plants? Have plants a sense of hearing? This is a difficult field for we still do not know enough to be sure of the answers. Plants require carbon dioxide (CO_2) for their metabolism, and we exhale CO_2 with every breath. Many reference books (particularly those for children) get the amount of atmospheric carbon dioxide wrong. The proportion of CO_2 in the atmosphere is vanishingly small. The air is about one-fifth oxygen and four-fifths nitrogen. What comes third, at 1 per cent? No, *not* carbon dioxide. The third most abundant gas in the air we breathe is argon, the gas used in electric lamps, a gas of which most people have never heard. There is three hundred times as much argon in the air as there is CO_2! Carbon dioxide is present at only one part in 3000 which makes it a very rare gas in the atmosphere.

For plants used to such a rarefied supply, humans breathing over them could be a godsend. There is 4 per cent CO_2 in the air we exhale, which is over a thousand times as much as there is in the air. That may help plants to grow. But there is also the question of the vibrations transmitted in speech. Plants respond to many mechanical stimuli (touching, stroking, wounding) and it could be that the attenuation of sound waves as they pass through plant tissues can trigger a hormonal response. This is pure conjecture, but to those who speak to their plants I say 'talk on'. The plants may benefit from the CO_2, they might even detect the vibrations of your voice, and a plant which is so intimately related to its grower will assuredly benefit from the conscientious care such a person could confer.

Some other current beliefs are less well-founded. There is a fashion for the production of crop plants that are resistant to pests and diseases. Rather than spray insecticides and fungicides onto plants, many people feel safer with a variety which resists attack 'naturally'. Some plant varieties resist insect invasion by having a covering of fine hairs which keep the insect pests at bay; others secrete compounds which prevent attack. These plants secrete poisonous molecules as a way of resisting attack in nature, and the resistant strains we use in horticulture and agriculture simply contain more of them. Cassava, for example, is a staple crop in tropical countries and survives for prolonged periods in the ground, where it forms a vital food reserve for populations otherwise liable to extreme shortages. Bitter cassava survives because it contains cyanide, and this poisons any would-be pests. In preparation for cooking, the cassava is grated and boiled, which drives off the cyanide radical and makes it safe to eat. In Western countries the product is known as tapioca.

It is important to realise that pest-resistant plants remain free of attack because they are rich in pesticides. Carefully used, sprays have much to commend them. Their use can be confined to at-risk situations, whereas plants that show biochemical resistance may contain hidden pesticide molecules. We understand the chemical composition and biological effects of sprays that are commercially available, whereas those produced internally by plants may be of unknown composition and unpredictable effects. Finally, the natural components of many plants pose hazards to animal life which we would do well to avoid.

Mad cows, sane plants

Plants have many other connections with humans and with the way we conduct our affairs. We have seen how they share our hormones, send out signals similar to our own nerve impulses, tire of repetition much as we do, and respond like us to the outside world. Have they a relationship to the major preoccupations of our era – mad cow disease, for example? Many of the chemical compounds produced by plants are used to defend themselves against attack from predators. Cyanides are widespread in natural plants. They are particularly abundant in brassicas (the cabbage family). During recent years the cultivation of oil-seed rape and the use of such brassicas as animal feed has greatly increased. Since the mid-1980s these cyanide-containing plants have become increasingly widely used in agriculture. At the same time, surplus cassava has also been used in Britain as cattle feed, and has even been recommended for use as a bulking agent in some hamburger products.

The harm brassicas can do is already well known. Sheep raised on meadow grass and turned out to eat brassicas are known to be liable to attacks of kale sickness, which is a response to the cyanide component of the kale. Harm from natural cyanide compounds has been recorded, too. Tropical ataxic neuropathy (TAN) is a disease of humans who rely on cassava as a major component of their food, and it is caused by the cyanides that remain. One of the post-mortem signs is a 'spongiose appearance in the white matter' of the brain, and a suggestion has even been made that there might be 'scrapie-like fibrils' in these specimens.

Here we have reports of a spongiform disease in animals fed on cyanide-containing plants. The use of these by-products has increased elsewhere in our diet. Their value as a bulking agent in food products has become more general in recent years, and it has been suggested that bovine spongiform encephalopathy (BSE) and even Creutzfeldt-Jakob disease could be the result not of infected cattle feed, but of the ingestion of excess cyanides in the diet. The incidence of TAN in people who consume large amounts of cassava already shows that similar diseases are well documented. To claim that plants caused the outbreak of BSE would be unjustified. If that were so, we would expect to find a close correlation between the extent of the usage of brassicas and cassava products and the occurrence of BSE. In fact, the increase in the use of these supplements has been international, whereas BSE has been concentrated in the UK. That alone provides evidence against such an association, but the facts remind us of the dangers of concluding that a new disease is understood, when there are other theories which could have been brought into the argument.

The lessons remain. Plants produce these poisonous chemicals to enable them to carry out their internal biochemistry, and to protect them against invasion from pests and diseases. Many of them we already harness for our own benefit. Digoxin and related drugs, for example, are still extracted from the foxglove *Digitalis lanata* and they doubtless perform important functions within that plant. The countless products of the plant kingdom which could be beneficial remain to be investigated. The size of the herbal medicine market is enormous, yet relatively little has been published on the efficacy of the agents themselves. It is often said that complementary medicine has a spiritual dimension which is not amenable to scientific examination. I doubt that. Early in 1997 I chaired a major meeting on complementary medicine, and said then what I believe to be the case:

> *Many present-day remedies arose from herbal medicines. They include the cardiac glycosides from* Digitalis, *curare originating in the Amazon basin, the salycilates found in willow. Any of them, if tested in a conventional controlled trial, would prove their efficacy. To imbue herbal remedies with a spiritual component is unnecessary: if they work in the body, the effects will show in scientific assessments. The herbal remedies of tomorrow can be tested, like the innumerable herbal remedies of the past. To pretend otherwise is to debase the discipline, and to confuse the patient.*

The air of mystery surrounding herbal medicine is hard to dispel, but not if we acknowledge the common principles that link us to the world of plants. It is not a chance event that a plant product has effects on humans. Any active compound in a plant may be equally active in an animal. We all share our common nature as living organisms; we need sense organs to find food, to locate water, to ensure that we maintain our inner environment in a suitable balance for the survival of our body cells. Plants are permanently in contact with the water they require, and can capture solar energy more efficiently than human science can imitate.

Meanwhile, is it true that plants can show emotions? Some researchers have claimed that a plant silently screams when uprooted or cut; others have shown that a plant wired up to a polygraph will send out signals if animals nearby are harmed. I would not be surprised if some of these demonstrations turned out to be true, for I have no doubt that we underestimate the capacity of plants to process information and their awareness of what is going on around them. But facts are facts, and most of the published accounts where scientists have tried to duplicate these demonstrations have drawn a blank. We do need to be able to repeat these observations, for repeatability is the key to good science. So far these claims have not been substantiated (though I rather hope that one day they may be).

Solving the pollen problem

My views on plants as communicators offer a conceptual answer to one prevailing problem: why do flowering plants have such exquisitely sculpted pollen grains, some of breathtaking complexity? Pollen grains are spores, and most spores have surfaces that are sculpted in a similar fashion. It would not be easy to identify fungus or fern spores on the basis of their appearance. But the pollen grains of flowering plants are different, and each is distinct.

What is different about pollen grains? Why should they be so varied, and so beautiful? If they were objects large enough to handle they would be collectors' items. Indeed, they already are collectors' items, for many microscopists have extensive collections (the grains are used to demonstrate microscopy to newcomers, as much as for reference purposes) and photographs of pollen grains regularly win awards. There is one on the front cover of a microscopical journal as I write; it arrived this morning, and caught the eye of everyone who glimpsed it from across the room. Why should a mere pollen grain exhibit such architectural complexity? There is a difference, of course, between pollen grains and spores. Spores germinate and from them grows a new plant (a fern prothallus, a new fungus colony, whatever). Pollen grains, when they germinate, send out a pollen tube which travels down to the ovum to fertilise the female nucleus within. This is the crucial difference: pollen grains are sex cells. As such, if a mismatched pollen grain from one family of plants were to grow on the pistil of another family, the fertilisation would abort. The point of pollination is to match the right pollen grain with the right pistil.

A system which relied on matching the cytoplasm or the chromosomes would be completely impractical. Every pistil would be crowded with pollen, each producing a pollen tube, and the conflict would be unmanageable. A flower whose pistil was crowded with competing pollen tubes could not function. What we need is a method of selecting the pollen grains so that they are detected when they arrive at the right site. Some flowers manage to guide pollen grains to their destination by using fertilisation methods which direct a pollinating insect from one flower to the next of the same species, but even that is not necessarily foolproof. There must be a process by which the incoming pollen is recognised. Plants cannot perceive an image and therefore need some other sense to take the place of vision. I think we may have the answer. I believe that the pollen grain profile is distinct from one species to the next as a means by which a host pistil recognises the identity of the newly arrived pollen. Like the chemical signals that pollen grains exchange with a host pistil, the sculpturing is equivalent to the brightly coloured feathers of a male bird. The bird sees the effect with its eyes, and plants perceive the sculpturing through contact with the pistil. If this is true, then the sculpturing is not merely a matter of beauty to the eye of the microscopist. It is the physical characteristic which marks out one grain from another, and allows plants to make sense of the crowds of pollen they might otherwise encounter.

We use the shape of pollen grains to distinguish one species from another. How percipient: flowers have already done this for millions of years.

Languages of plants

Plants have a vast vocabulary of signals and responses. They can detect the signs of a change in their environment and adjust their metabolism to anticipate its effects. They have finely tuned senses for signs of environmental stress which can compensate for its effects. Plants are able to detect the signals that herald a change in the leafy canopy above them, and alter their architecture in order to retain their fair share of light. Plants can detect changes in the total amount of light reaching them, and can also sense changes in the spectrum of the light. The most obvious method is for the plant to respond to the amount of light energy, sometimes even by twisting the leaf to avoid too much light, and the extent of photosynthesis within the leaf. However, we also believe that plants have specific light sensors which act as 'eyes'. They detect the nature and extent of the light, but consume very little of its energy.

This may be regarded as a sense of 'sight', quite distinct from a mere response to the effects of light energy on the rate of metabolism of the plant cells. Plants detect light in colour. They have sensors specifically for ultraviolet, blue, red and far-red radiation. These sensors provide them with a rich and detailed impression of their situation. Plants use the ratio of red to far-red radiation to control their rate of growth. By changing the speed at which stems elongate, a plant seems to be able to avoid future deleterious effects on its development. In particular, changes in light quality can forewarn the plant about future changes before any shading by neighbours has occurred, and we have seen that plants detect wind, too. Changes in calcium ions, Ca^{+2}, trigger alterations in gene expression which modify the way the plant grows. The metabolism, the expression of genes and the rate of growth of the plant are an integrated response to all this sensory input.

Meanwhile, plant roots adjust to the availability of nutriment, and favour areas where nourishment is most abundant. Their sensing of the availability of food and water allows the plant to adjust its rate of nutriment uptake. Not only can they sense gradients of moisture in the soil, but they change their rate of growth in the presence of nearby roots. It seems that they are able to control their growth in order to avoid too much competition for scarce supplies of raw materials. There are mycorrhizal associations between plants and fungi, in which a fungus colonises the roots of a host plant (sometimes even penetrating inside the host plant cells, but causing no disease). These relationships bring a number of benefits. Not only do the fungi process wastes in the soil and recycle them as foodstuffs for the growing plant, but the interchange of compounds between fungus and plant provides many opportunities for communication and the transmission of warning signals.

Plants have many of the senses possessed by humans. They have sight, as far as they need it; they have a sense of touch (sometimes to an extraordinary degree); can sense temperature, and – through gravity – they can tell 'up' from 'down', or 'left' from 'right'. Twiners can (usually) tell clockwise from anticlockwise. Plants can remember stimuli, and tell one form of stimulus from another. They can communicate, and they cooperate to survive. If plants required more intelligence, they would have developed it. As it is, their senses and the limits of their sentience are exactly what they require. Some of the senses in the plant world are already more highly developed than ours (the sense of touch, for example). No longer should science regard a green plant as a simple organism which endures what it must, and adjusts like a chemical system. We owe plants respect, for on green plants we all rely for survival. They are not our subjects; plants are our cousins.

6

Senses among the Oldest Forms of Life

Microbes were the first organisms to appear on the new earth. Most are invisible to the unaided human eye, yet they can sense where they are going and they can recognise friend and foe. Many boast senses superior to ours, and some contain sense organelles of breathtaking complexity. One microbe has a complete and fully functioning eye, for example, within its single cell. It is not true that you cannot see microbes with the naked eye. Most cells are indeed visible. You can see amoebae, for example, even though they look like nothing more than a pale pink dot smaller than a full-stop. Most bacteria are beyond the human eye, but typical protozoa, algae and fungal cells can just be discerned.

The classification of microbes is a perpetual source of debate among taxonomists. They create, and re-create, classes and families, species and genera, as though this were the whole aim of science. Classification does not matter to the other organisms. They know who they are, and that's what counts. We need to find the right way to interrogate organisms about their nature, rather than impose arbitrary conventions on them. Microbes are organisms which need to be magnified for scrutiny. Most are single-celled, but some of them (including some microscopically small nematode worms, and the spectacular rotifers which are related to them) are made of many cells, but are the same size as some single-celled organisms. Some single cells are much larger than you might expect. An ostrich's egg begins life as a single cell. The Cauperlaceae, a family of strange algae, grow as one huge cell which can be many metres long. A recent discovery in sturgeon, the fish from which caviar is obtained, has been of a giant single-celled animal organism several millimetres in length. Most animal cells are hundreds of times smaller. In biology, there are always striking exceptions to our handy rules.

In this book I will ignore the viruses since they are not organisms in any sense, and are hard to envisage as 'living' at all. Viruses are like rogue genes which spread from cell to cell. They do not reproduce, but undergo a process we call *replication*. The virus commandeers the host cell, and switches its operating instructions to make more virus. That's hijacking, not living. Other infective agents (like the infective principle of bovine spongiform encephalopathy and Creutzfeldt-Jakob disease) are believed to be prion proteins which do not even contain DNA or RNA. These are a new form of infection, simpler than viruses on the scale of structural complexity.

In recent years there has been a great vogue for reclassifying the main groups of living organisms, with different terms being introduced. The single-celled organisms fall into two main camps, the prokaryocytes (those with simple cells and without a separate nucleus) and the eukaryocytes (organisms with organised cells containing nuclei, mitochondria and the other organelles seen in multicellular forms of life). The single-celled prokaryocyte organisms are often known as monists or monera, while the eukaryocytic types are known as protists or even protoctista. For all the increasing proliferation of the taxonomic terms, the organisms remain the same. Here is a description of the main types of microbes, allocated to their familiar groups.

Bacteria

Mention bacteria and most people think of disease germs. Many of the major epidemic diseases were caused by bacteria: typhoid and cholera, anthrax and tuberculosis. As in human societies, the proportion of 'bad' bacteria is small compared to the 'good' ones. We know the bad ones as *pathogens* (from the Greek, meaning 'disease-causing'). In an article for the scientific journal *Nature* I coined the term *salugens* – literally, 'health-producers' – for the countless types which convey health to larger organisms. Examples would be the bacteria found within the cells of some insects, which remain stunted without their bacterial partners. Others are the bacteria that infect the roots of plants, and allow them to change nitrogen gas from the air into usable nitrate for growth. Bacteria recycle countless components in the environment, and can consume compounds like cyanide which are hazardous to humans. Any idea that bacteria have a language seems absurd. They are simple prokaryocytes, without a proper nucleus and with the most primitive type of cell organisation. Some can form spores which are unusually resistant to heat and desiccation. Most bacteria usually prefer an alkaline environment. They ordinarily shun bright light, though some contain photo-receptors and can capture solar energy. The blue-green algae are often included in this group, because they are prokaryocytes too.

Bacteria are typically a few thousandths of a millimetre in size. They are customarily measured in μm, known as the micrometre, where 1 μm is one-millionth of a metre. Human cells average about 15 μm across, and bacteria are more like 2–3 μm. They vary considerably, however, and one species which was discovered in 1985 is visible to the naked eye. It is

Epulopiscium fishelsoni, and was found in the intestine of the brown surgeon-fish *Acanthurus nigrofuscus,* which inhabits the Red Sea. Little is known of this novel organism. It is large – up to 600 μm (over half a millimetre) long and 80 μm in breadth – about the size of a whisker cut from a man's chin during the morning shave. This bacterium will so far only grow inside the host fish, so studies of its behaviour are proving difficult.

There are many senses in bacteria, including their own version of taste, and some can sense a magnetic field through particles of magnetite in the cell. In this way they can swim along lines of force in the earth's magnetic field in temperate latitudes, and so navigate down through layers of mud to safety. Many bacteria exist in cooperation with other organisms, and adapt their behaviour in the light of what they sense of their surroundings. The passage of a newborn baby through the birth canal inoculates its skin with communities of bacteria from the vagina. Many of these will be salugens, which help protect the baby from harmful germs it may encounter in the outside world.

The genus *Rhizobium* is vital for agriculture. These bacteria can sense the roots of newly germinated leguminous plants (beans, peas and other pulses) which they proceed to infect. The result of the colonisation is the formation of nodular growths. These root nodules look like the sign of some kind of disease. In fact they are vital for the plant's health, because the bacteria that inhabit the nodules can capture nitrogen from the air to make nitrates which the plant can use to create new protein. For centuries, farmers have known that it pays to allow clover to grow in a field every few years. Clover, the name for several species of *Trifolium,* has these root nodules and, if left to grow for a season, will replenish the soil with nitrogen. This was a method of enriching soil before the era of chemical fertilisers. Organic farming utilises this process.

One great problem in all farming is the removal of crop

material from the field. In nature, waste vegetation is recycled to feed next year's crops. Farming removes the crop altogether, so each year that goes by sees a reduction in soil quality. The function of soil bacteria is to create new nitrogenous materials from the air – and though our manufacturing processes give us a chance to create high-efficiency farms, it is worth noting that the smallest living organisms known to science are able to make nitrate from air. Bacteria are far cheaper than factories, and require no input of external energy. As a source of new amino-acids, bacteria are marvellously miniaturised production units.

Some bacteria form patterns when they grow, which demonstrates that they may well have a sense of position or orientation. Others recognise one another and move together to form communities. *Chondromyces* are organisms which are free-living in soil. From time to time they enter a reproductive phase. When this happens they find their way together and form a rounded body (quite like the slime-moulds described on p. 240) which begins to behave like a single organism. The cells climb up over each other producing a branched aerial structure, at the ends of which the cells form spores. The spores develop a resistant cell wall, so they can be spread by the wind and germinate in some far-off habitat. Other bacteria have the ability to select members of the same species which are of a different gender. The two strains (traditionally known as plus and minus rather than 'male' and 'female') send signals to each other and, if they recognise sexual compatibility, they will swim together and fuse. A small tube projects from the plus or 'male' cell and joins with the minus or 'female' recipient. Through this tube, genetic material can pass. In this way, cells of many bacteria (like *Escherichia coli*, which lives in huge numbers in the human intestine) can pass genes from one to another. It is this transference of genetic characteristics which gives rise to new pathogenic strains, like the *E. coli* type O157 that causes severe dysentery and is becoming increasingly widespread in modern beef cattle. This strain acquired two genes from a dysentery bacterium sometime in 1982, and has been spreading ever since. The result of this genetic change is that a normally harmless form of *E. coli* has the additional ability to produce a powerful toxin which can destroy tissues in the body. The resulting disease is a singularly unpleasant cause of death.

Sexual reproduction is common in bacteria. Many microbes are known to be able to pass on their ability to resist antibiotics. In this way, resistance can be transferred to new organisms. In the ordinary way, *E. coli* is nothing to fear. It is the commonest bacterium in the human body, and normally does us no harm at all. Because of its abundance in the human intestinal tract, *E. coli* has been studied intensively for many years. We know it relates to other cells, because it can recognise its opposite number and undergo conjugation, and also because colonies can form patterns. If cultures of *E. coli* are grown in a medium containing chemicals involved in the cell's biochemistry, they will swim together in predictable patterns. Compounds like succinate or fumarate will stimulate them to seek each other out and swim in coordinated arrays. This truly is synchronised swimming on a microscopical scale. *E. coli* moves fast, swimming up to twenty times its body length per second (equivalent to a human six feet tall swimming as fast as a power-boat). In different chemical environments, bacteria form differing patterns. Some strains of *E. coli* lack the ability to sense specific compounds. One is 'blind' to the amino acid serine. These cells form beautiful star-shaped patterns when grown in culture. Twenty years ago I wrote of the way microbes communicated, and the idea of a language then seemed far-fetched. It is still a strange idea, but at last we are gaining a greater understanding of the way communication could operate. Molecular biology is shedding light on cell communication, and this is one of the greatest contributions of reductionist science to the holistic endeavour of organismal biology.

For many decades we have known of the *bacteriocins*, chemical compounds that bacteria can direct at opponents to destroy them. They attach to the enemy cell, break through its cell membrane and destroy it. At one time it seemed that there was a single form of bacteriocin, but we now know that there are many of them, each adapted for a specific purpose. Bacteria use these compounds to establish a territory. If a mixed community of organisms is established, they maintain their position by eradicating outsiders. Monitoring the community allows us to follow which ones survive, and which ones die. It is always the bacteria lacking effective bacteriocins that disappear from the group. Looking for chemicals which destroy other cells is a legacy of our twentieth-century fixation on violence, conquering, and eradication of an enemy. Bacteria actually send out far more messages which are mere signals for their neighbours. These enable them to spread out and avoid competing for scarce nutriments, for instance. Other signals lead to the grouping of bacteria in colonies. We have seen how this happens in the case of *Chondromyces*, but this example is well known because its effects are clearly visible. In the real world, all bacteria are communicating in this way. The common sausage-shaped bacterium *Bacillus subtilis* was one of the first I ever cultured. It has a remarkable phase in its reproduction, for a single cell sometimes divides to produce daughter-cells of different sizes. When this occurs, the larger cell engulfs the smaller, and the resulting cell then secretes a hard cell wall to become a spore. No longer is it enough to regard bacteria as simple factories which split in half.

Although we still know little about the organelles of sense within bacteria, we do know that *E. coli* has the ability to sense desirable compounds, and to avoid undesirable chemicals: but how? To identify the bacterial nose, antibodies were labelled with microscopic particles of colloidal gold. These were attracted to the organelle of sense, and the gold later revealed itself as dense particles in an electron micrograph. Surprisingly, the particles were collected together at one end of the cell. We had always assumed that bacteria had fairly diffuse organelles of sense, and detected a chemical in the environment by swimming up or down a gradient of concentration. Now we know the simpler truth: the *E. coli* cell contains its own nose. Some of the proteins involved in the sense of smell have been identified, and so far we know that proteins denoted by CheA and CheW trigger a chemical cascade which allows the cell to smell its surroundings. As in humans, the nose of *E. coli* is found at the front end. Even the senses of such ancient organisms as bacteria are far more highly refined than we have suspected.

Fungi

Fungi are great agents of recycling. They can take dead matter and digest it. Their process of growth means that they are, in effect, changing waste matter into fungus protein. Fungi typically grow as long fine threads which, *en masse*, form a mycelium looking a little like damp cotton wool. Many of the microscopic fungi produce spores at the ends of these threads, but some larger fruiting bodies form when masses of the threads (hyphae) grow together to form a solid body. A mushroom is one of these. In this case, the spores are reproduced from the fruiting body itself – but these fungi, being easily visible to the naked eye, are hardly microbes. Fungi prefer an acidic environment and rarely grow in bright light. Among the chemicals which comprise the cell walls of fungi is chitin, another form of which constitutes the outer carapace of insects. Fungi are not normally found in the sea. Nor are insects, come to that.

Fungi often cooperate with other forms of life. The most

intimate example with which we are familiar are the lichens. These can grow on barren rocky surfaces, for they are colonies of closely intertwined fungal hyphae associated with algae. The fungi recycle chemical foodstuffs and provide the structure of the plant, while the algae capture sunlight to provide energy. Other associations are less well known. For instance, plant communities rely heavily on fungi. The roots of forest trees support growths of fungi, and the toadstools we see in woodland during the autumn are the fruiting bodies of these colonies. There are orders restricting the gathering of many of them. This is not as sensible as it seems. Although it is excellent to allow fruiting bodies to mature (and therefore to shed their spores to help the species survive), the toadstool fungus is only a minute part of the plant. Most of the fungus is a mass of threads hidden in the soil, and closely interlaced around the roots. In some cases the fungi enter the cells of the roots, for all the world like an infection. Instead of harming the host plant, they provide freshly recycled chemicals for the plant's roots. The association between the root and the fungus is known as a *mycorrhiza*.*

Many plants seek out a fungal partner when their seeds germinate. Orchid seeds, for example, often need the presence of the fungal partner before they will successfully germinate. The fungus and the seed communicate to each other and coordinate their development together. The way fungi grow together to form fruiting bodies – toadstools that are instantly recognisable by their shape and colour – shows how complex their language is. These organisms exist as free-growing fine threads forming a felty growth through the soil. Growing together to form a highly organised toadstool calls for coordination and timing of a sophisticated kind. To do this requires a complex and refined

---

* Many reference books describe a mycorrhiza as 'a fungus', but that is incorrect; the term describes the association between the fungus and the root.

language. We still know little of it, though we can marvel at its effects.

Algae

Like the fungi, algae exist in microscopic and larger forms. The larger forms include seaweeds, some of which can grow many metres in length. Microscopic algae are everywhere. You can find them in snow-fields on high mountains or out in the open ocean. The typical algal cell contains chlorophyll, or some other photosynthetic pigment. Most have a cellulose cell wall, but some (the diatoms, for instance) build a glassy skeleton of pure silica. Some microscopic algae have a flexible cell wall and swim with the aid of a projecting whip-like flagellum. Zoologists claim that these are really animal organisms, but they contain chlorophyll and botanists are usually happy to regard them as plants. Bacteriologists, on the other hand, often claim that the tiny blue-green algae are really bacteria. Microscopic algae are common fresh-water organisms, and they respond to light and ambient moisture with a range of adaptive mechanisms. Although we know that many of them contain complex eye-spots, little is known of the other senses that algae must possess.

Protozoa

These are microscopic animals. Most are just visible to the naked eye, and some can be seen actively swimming if you look closely at a phial of pond-water. The best known must be amoeba, a shapeless single-celled creature which lives on the sur-face of mud. Others swim actively by means of a single flagellum, or a coating of fine cilia across the surface of the cell which beat in regular motion like wind blowing across a field of wheat. Whereas most algae are content to stay put and synthe-sise new materials with the aid of sunlight, protozoa are almost always on the move. They have many senses, including vision.

Some are capable of constructing shells made from sand-grains cemented together, and most of them mate. This in itself demands a range of senses, of which some sense of enjoyment must be a feature. An organism grazing on its food will not turn to sex unless it has some means of detecting that it prefers one activity to the other.

The essential simplicity of microbes is largely an illusion. We say that an amoeba is shapeless, but that's not quite true. No two amoebae are identical. The same amoeba, on any two separate occasions, will look different and can never look exactly the same again. But there is a congruity to the shape; we can tell one species of amoeba from another because of the proportions of the cell and the shape of its projections. The way the amoeba is taught to us is as a shapeless blob of protoplasm that simply crawls around in water all day. *Shapeless?* It has a shape of its own which is perfectly recognisable. It cannot be 'any shape at all', for it will never be a square or star-shaped. Nor can it be the shape associated with a morphologically different species of amoeba. *Crawls around?* This is an organism with a sense of purpose. Trap it in a cul-de-sac on a glass slide under the microscope and the movement of the granules within the cell suggests that it turns round before finding its way out: it has a head and a tail. Present it with a food particle and it will inspect it. If it approves of what it finds it will flow round the particle and engulf it, taking it into its cell body to be digested. This cell is made of water-soluble constituents, and yet it lives in water. That does not prevent it from maintaining its shape and its internal composition. One remarkable thing about an amoeba, a water-soluble system, is that it does not simply dissolve in the surrounding water and vanish.

We even have amoebae within us, and we depend on them for our survival. These amoebae are the white cells of the human bloodstream. Some produce antibodies to inactivate an invading pathogen. Many of them act much like the amoebae on a pond: they crawl around looking for prey. If they detect an invading bacterium in the neighbourhood – a staphylococcus or a strepto-coccus, say – they can congregate where the bacteria are found and ingest them. Consider what this means: the white cells have the ability to detect an invader, to travel to meet it, to identify it as hostile, to inactivate it, eat it and destroy it. This reveals a considerable degree of sensory and motor coordination. The amoeboid white blood cell knows where it is going and what it has to do. It is not controlled by any hormone or messenger system in our bodies, for it is entirely self-propelled. We depend for protection from illness on cells much like the humble amoeba which lives in the gutter. Amoebae have a set of finely tuned senses on which they depend to find their way to food, identify it and eat what they need to survive. We have under-estimated their capacities for far too long.

I think this could explain the outbreaks of Legionnaires' disease. The causative organism is a tiny bacterium named *Legionella*. It normally exists as a parasite of the amoebae which live in warm water. Once in a while it causes an outbreak in humans, often through contamination of an air-conditioning system or a public shower. My guess is that the *Legionella* organisms mistake the white cells of the human bloodstream for the amoebae in which they normally exist. To the germ, one amoeba is much like any other. They think they are colonising a free-living cell, while they are actually making us ill. The *Legionella* organisms are not entirely wrong, however; a human white cell is an independent organism. It certainly is not under the control of the human in whom it lives.

The 'simple' amoeba can do everything it needs to do. It exists, free-living in a pond. In another guise, it is part of your body, searching out germs that might bring you down. Truly, this is a pinnacle of success, and not a lowly creature. Amoebae can do some things which we cannot. Some would be well worth emulating: thus, an amoeba adjusts its rate of reproduction to the available food supply, and never over-eats in conditions of abundance. Others are just a dream: if their environment is threatened, many amoebae can build a sealed, protective capsule which will allow them to survive for prolonged periods without food and water. That would be a great feat for a human.

There are even amoebae that construct homes for themselves. They find fragments of sand or stone and use them to build a delicate enclosure. The architecture of this artificial shell is of great beauty and the work is done with predictable precision. Each shell-building amoeba has its own trademark. The shape of the shell is recognisable from one species to another, and it is used by scientists to identify which is which. The simplest amoeba can find its food, move from place to place and avoid unpleasant environments. The shell-building species are exhibiting the signs of many of the abilities on which we pride ourselves. Peoples of the past created monuments of stone which we take as arbiters of their great abilities in times long gone. Present-day amoebae are building stone homes of their own. We can only speculate on the senses they need to do that. If humans need intelligence, wisdom and insight to make dry-stone walls, an amoeba cannot be entirely devoid of data-processing abilities if it is to make a home for itself, and to know when danger approaches.

Other microbes

There are many rotifers and tiny nematode worms in fresh water and soil. Some nematodes are much larger, like the giant parasitic worms that infect pigs and can be larger than an earthworm. The threadworms which infect infants are nematodes. Most are much smaller, including the eelworms which infect crops and garden plants. There are more nematodes in the world than any other multicellular organism. They abound in the soil, and the sand of the sea-shore is bursting with millions of them; indeed, it has been said that if you were to remove the minerals from our planet, and all other forms of life, the mass of nematodes would still preserve the shape of the earth. How they interact and detect their prey is not known. Indeed, it is likely that among the organisms we call nematodes are some still-undiscovered groups of living organisms. Microscopic nematodes and rotifers all share one unusual characteristic: each adult member of the same species is composed of the same number of cells, and each organ is built of the same number of cells.

Rotifers propel themselves by means of leech-like crawling movements, and also by paired swimming organs bearing circular arrays of cilia. They have the appearance of paddle-wheels whirling around. To early research workers these looked like wheels. In recent years, evolutionary biologists have even posed the question: why didn't nature evolve wheels? She did. There are tiny rotors within the basal structures of the flagella which propel bacteria through their watery environment. Mechanical wheels on cars and bikes are produced by humans, and – if we regard *Homo sapiens* as the pinnacle of natural processes – they are therefore a product of nature. Our legs are the organic version of wheels, with 'tyres' (the soles) which are continually renewed from within. Mechanical wheels are crude by comparison. They need periodic servicing and the old tyres have to be replaced by hand. Wheeled vehicles cannot walk upstairs or climb trees, so they have marked limitations.

Fungi and light

Sight is the most sophisticated of the senses. We pride ourselves on our ability to detect small changes of mood in our fellows, to spot if someone lies, to recognise birds at a distance or to marvel at the stars; all by using our eyes. Humans look patronisingly at animals, knowing that dogs and cats cannot see the colours we see, or perceive the details on which we rely. We know that pigeons can be trained to pick out misshapen pills on a conveyor belt, so we know that some other creatures can use their eyes, but we pride ourselves on our uniqueness.

Microbes have eyes. Single cells can see. Nobody has found an organ of sight in a bacterium, for a prokaryocytic cell lacks the basic components to do the task. But many eukaryocytes have organs of sight. There are even eyes in single-celled algae and fungi. Light has a crucial effect on the liberation of spores in many types of fungi. If ripe cultures of the fungus *Sordaria* are kept in the dark, they release only small numbers of spores. When they are exposed to light for 12 hours at a time the rate at which spores are released increases dramatically. Interestingly, the rate of spore release peaks halfway through the daylight period, and drops back again by the time darkness supervenes. It is as though they have some internal sense of timing.

A sceptic will say that yes, well, it is clear that light helps spores to mature in some way. Fungi need to release spores during the dry hours of daylight, to maximise dispersal by air. There must be some simple relationship at work. Logical as that may seem, fungi are wiser than that. There are other fungi, including *Apiosordaria verruculosa*, for which you have to wait eight or ten hours after exposure to light for the maximum spore release to peak. This fungus has a regular pattern of spore release, at its minimum at the start of a 12-hour period of daylight, and peaking to a maximum shortly before darkness comes on again. Experiments show that, if it is then kept completely in permanent darkness, the same 12-hour cycle continues. These fungi can keep time, even when outside cues are eliminated.

In other groups, including the fungi *Hypoxylon* and *Xylaria*, light has the effect of directly inhibiting the release of spores. Even a small exposure to low-level light stops the release of spores. *Daldinia* cultures, kept in the darkness for prolonged periods, retain their 12-hour periodicity for days on end. Exposure to 12-hour bursts of light reduces spore release almost to zero, and this pattern persists in conditions of total darkness for a week or more thereafter. These fungi possess a biological clock. The effects depend on temperature, too. If *Sphaerobolus* is grown at 20°C (68°F), it releases its spores during the light periods; when grown at 10°C (50°F) this behaviour reverses, and spores are released during the hours of darkness. Fungi have developed to do what is best for themselves in the circumstances.

In some fungi there are the beginnings of an actual eye. *Pilobus* produces its spores at the end of a single hypha, as do many simple fungi. The pin-mould you can see on stale bread does this too. The sporangium forms a cushion-shaped capsule at the tip of the thread. Below the sporangium, the hypha swells out to produce a round, clear globe. It acts as a lens, focusing light towards its base. At the lower end of this lens-like swelling, just where the hypha begins, is a ring of cytoplasm which is rich in carotene. Onto this the light is focused. If the light shines from one side, the fungus adjusts the position of the hyphae so that they point directly towards the light. The image of the light-source is focused onto the carotene light-receptor, and if it is unevenly distributed, the fungus adjusts its position until the light is evenly focused. This is the essential structure of a human eye: a focusing component, acting as a lens, a light-receiving carotene ring, acting as a retina, and there are clearly the means to detect any unevenness and coordinate a response. All this, remember, is in a single strand of a fungus.

Pilobus lives on cow-dung, and uses its 'eye' for more than merely aiming its consignment of spores. The lens-like capsule swells out as it develops, until the time is reached when it can contain its internal pressures no longer. At this point, the capsule bursts, shooting off the spores from the sporangium at up to 50 km/h (about 30 mph). There is one nematode worm which uses this for its own transportation system, for it slips into the mass of spores and waits there for the capsule to burst. As the spores are ejected, the nematode travels with them. In this way it manages to spread to new areas which otherwise it would find hard to reach.

Fungi as carnivores

The ability to trap animals is found widely through the plant world. Not only is it done by highly evolved green plants, but there are even microscopic fungi which set traps for animals. There are thousands of them in a single spoonful of earth or compost. These are the predatory fungi. They exist not merely by the decomposition of animal remains, but by springing a trap which seizes the struggling prey and kills it. They have been known for over a century. A fungus trapping an animal was first recorded in 1888. Tiny nematode worms are the most frequent prey of these carnivorous microscopic plants, but other minute animals are trapped by them, too. I have seen them capture rotifers, for example, and there are reports that they can even trap an amoeba.

Sticky spores are the simplest means by which fungi infect tiny creatures in the soil. As the spores form, they develop a coating which sticks fast to passing microbes. Sometimes a passing organism even eats one of these spores. In time the spore hatches out. From within grows a fine transparent hypha which enters the host's body. As the microbe withers, the new fungus grows, and when the last of the prey is consumed it enters its reproductive phase. Masses of tiny spores are produced which become coated with glue. The cycle is then ready to start anew, just as soon as some hapless tiny creature stumbles across the scene.

Different fungi trap their prey in different ways. Some use an adhesive to catch their prey, rather than to stick to the exterior. These fungi ordinarily exist as conventional saprophytic organisms, digesting organic wastes in the soil and utilising them as a source of both food and energy. However, they also secrete a sticky substance which coats the outside. Amoebae sometimes find they are caught by the secretions. If a nematode worm swims into it, it finds itself trapped. Further wriggling ensnares it in further loops of the thread-like hyphae of the fungus, until it is stuck fast. The effort soon exhausts the tiny worm and it dies, locked in the embrace of its sticky captor. The fungus sends out fine hyphae which grow through the body of the prey, digesting it from inside and feeding the fungus which caught it. High-powered optical microscopes reveal much activity within the fungal cells. As the prey is caught, tiny granules within the hyphae stream towards the point of capture.

Other fungi capture rotifers. They live in fresh water and produce tiny projections from their hyphae as they grow. A browsing rotifer, seeing one, assumes it is a tasty snack and ingests it. Once it has taken it into its mouth it is trapped. The hypha is like a fish-hook and the rotifer cannot break free. Fine branches from the hypha spread through the rotifer's body, digesting it and feeding it to the parent fungus. As the rotifer shrinks on its branch, the rest of the fungus grows larger with what it has consumed. The most dramatic of these microscopic hunters are the predatory fungi which lasso their prey. Under normal circumstances the fungi grow as conventional saprophytes. If there is a sudden upsurge in the number of nematode worms, however, the fungus changes its ways. As soon as it senses the worms it starts to sprout tiny loops from its hyphae. Each one is composed of three cells joined end to end. A nematode swimming by is likely to head for the loop, as it is an open aperture worth exploring in a cramped environment. As it swims through the cells of which the loop is composed they suddenly dilate, squeezing the gap down to a fraction of its size, and nipping the nematode firmly in its grip.

It is trapped. Sensing a problem, the nematode wriggles furiously to try to release itself. The attempts are almost always fruitless. Within one-tenth of a second the cells of the loop have blown up like an air-bag in a car. The nematode usually dies of exhaustion, and if it doesn't, the slowly growing hyphae that spread through its body will kill and digest it. Once in a while a particularly strong nematode breaks free and escapes. It is a short-lived victory. Fragments of the snare are always left stuck to the exterior of the worm. In time these grow to form new hyphae which penetrate the body and start to feed on the dying worm.

Yes, these are 'only' fungi. They may be such diminutive organisms that you have never even noticed their presence on the compost-heap. But they have a range of ingenious traps for the unwary creatures who pass by, and if you search within them you may see the hive of activity that accompanies each kill. There is more going on inside the fungus world than readily meets the human eye.

Microbial eyes

Many free-swimming protozoa have eye-spots. One of the best-known protozoa is *Euglena*, many species of which are found in fresh-water environments and also in the sea. A typical feature of *Euglena* is chlorophyll, located in chloroplasts, which capture solar energy and help to supply the cell with food. For this reason it is usually described as an alga, but we shall group it with the protozoa here because it has so many features in common with other distinctly animal-like organisms. From the view-point of *Euglena*, it doesn't matter where it is placed. These conventions are created by people, and the only effect it has here is where the subject occurs in my book.

Euglena has a flexible body which can twist and change shape as it moves. The organism can creep through small spaces, rather like an amoeba in a flexible envelope. For much of the time it swims actively, at a speed which exceeds that of an Olympic swimmer, size for size. It is propelled at speed by means of a single flagellum, a whip-like thread from the front end which it thrashes through the water. The waves of movement rotate the cell as it swims, and propel it steadily forward. Near the base of the flagellum lies the stigma, or (more accurately) the eye-spot. This is often dismissed as a simple pigment granule, but the electron microscope reveals a more intriguing structure by far. The eye-spot of *Euglena* is composed of two principal parts. There is a cup like a tiny retina, made up of about forty little spheres of pigment. Inside these spheres carotene has been identified, along with several other photo-receptor compounds. Lying adjacent to it is the paraflagellar body, which is shaped like a lens and has a crystalline structure.

The scientific accounts of *Euglena* orienting itself refer to these structures as though they were components in a primitive electronic device. If the stigma shades the paraflagellar body, they say, the cell automatically rotates until shading no longer occurs. It is like a simple photoelectric cell strapped to a tricycle. Imagine you are suddenly transported to a dark room, and you turn to look out of the window. I can imagine one of these scientists describing your responses: 'If the lens of the human organism is in shade, the body rotates until shading no longer occurs.' How patronising. You could say, instead, that the *Euglena* turns round until it can see where it is going. I am aware of the anthropomorphic trap that awaits people who say things like that, but the *Euglena* eye-spot does indeed resemble the human eye. There is a crystalline lens, shaped just as a lens should be shaped, and a cup of optical pigment granules shaped like a retina. I am imposing no outside, human-centred values on this description; rather, I am allowing *Euglena* the dignity of having its intricate beauty fully recognised.

There are eye-spots in some of the tiniest of all free-swimming organisms, and they are commonly found in algae which need to find their way towards the light. It should not be assumed that the eye-spot draws the cell inexorably towards a source of illumination, for if the intensity is too bright it will turn away and swim off to a place where the light levels are more tolerable. In these organisms we can see a variety of lenses and a range of retinal cups, though the basic pattern of a crystalline lens and an array of tiny spheres containing light-sensitive pigment is common to most. These are not mere eye-*spots* – they really are tiny eyes. In their most complex forms (in some of the dinoflagellates, infamous for causing poisonous blooms which can kill cattle) these structures are known as ocelli (literally, 'little eyes') and their optical design is unmistakable. The complexity of structure is a close parallel to that of the eyes of humans.

One ingenious experiment allowed scientists to restore the sense of vision to a 'blind' microbe. *Chlamydomonas* is a tiny, free-swimming alga. It is grouped as an alga, not a protozoan like *Euglena*, because its cell wall is more rigid and plant-like. *Chlamydomonas* swims by means of a pair of flagella projecting from the front of the cell, with which it rows itself along at speed. To one side of the cell is an eye-spot, with a pigment cup and a tiny lens. With this, *Chlamydomonas* can detect light and maintain an adequate supply of solar energy. One mutant strain of *Chlamydomonas* lacks the pigments essential for its sense of sight. Part of the structure is present, but not enough to create a compound that is sensitive to light. As a result, this mutant strain is blind, and swims without reference to the direction of illumination. In an experiment, rhodopsin, a key mammalian visual pigment, was administered to cultures of the alga. This triggered the restoration of full activity to the tiny eye-spot: the *Chlamydomonas* regained its ability to detect light, and swam as did its 'sighted' relatives. This is a crucial observation. Restoring the visual activity of the eye-spot through this procedure is gratifying, but the result is not unexpected. The important point is that the visual pigment rhodopsin came from animal origins, and not from plants. Here too we are reminded of the universality of the biochemistry of life, and of the senses we all share, even with our most diminutive neighbours.

An ability to detect light is found in many microbes, and in some the mechanisms of photo-detection have yet to be unravelled. The common amoeba, *Amoeba proteus*, crawls away from light. No light-sensitive pigments have been found, and there is no eye-spot, yet the response is unmistakable. The reaction is most marked for blue light, the red end of the spectrum having little effect. Shining tiny, focused beams of blue light onto different parts of an amoeba shows where the sensation is detected: it is just behind the front of an advancing part of the cell, or pseudopodium. If light is focused onto the outer layer itself, the amoeba takes no notice; nor does it respond to light shone on the centre of the cell.

However, the interior of the cell is filled with endoplasm, a semi-liquid cytoplasm in which the nucleus floats, along with a host of other tiny particles. The advancing front of this endoplasm is where the detection of light takes place. In some of the tiny mitochondria within the cell (the sausage-shaped bodies where the release of energy is centred) riboflavin has been detected, and it may be that an amoeba has minute light-sensitive granules which we have yet to discover.

One of the most spectacular of the swimming protozoa is

Stentor, shaped like a trumpet, bluish or lilac in colour, and clearly visible to the naked eye. It looks like a minute comma suspended in water, and it swims by means of a covering of beating cilia. Some species of *Stentor* contain tiny algal cells, with which they co-exist: the algae provide a food source for the cell, and the *Stentor* offers a home for the algae. This co-existence, symbiosis, is common in biological systems and may explain how complex forms of life first evolved. Within *Stentor* cells without these symbionts, photo-receptor pigments are common. If the cell swims towards light, the cilia stop beating backwards and suddenly reverse the direction of thrust. The cell stops swimming, backs up a short distance, then sets off away from the direction of the light.

Light is anathema to *Stentor*, and it has even been shown that a burst of bright light is enough to kill a *Stentor* cell. Some exceedingly delicate experiments have shown that you can measure different types of response by inserting microscopically small electrodes into the *Stentor* cell. Most significant is an action potential, which fires when light reaches a threatening level. This causes immediate contraction of the body from its extended trumpet-shaped configuration into a rounded ball. It is a protection response. There are also graded potentials, which seem to mark the animal's response to different shades of light. In free-swimming *Stentor* cells there is a third type of electrical response related to light bright enough to trigger the avoiding reaction. These are regenerative responses measuring about 50 millivolts, and they switch the beat of cilia into emergency reverse thrust. If a *Stentor* cell is transferred to a weak solution of caffeine, the photo-receptor chemicals are bleached out, and in this state the response to light disappears.

Blepharisma is related to *Stentor*. It is similar in size, but lacks the open trumpet mouth. Normal *Blepharisma* cells are pink because of their photo-receptor pigments. Well-pigmented individual cells can be killed by a burst of bright light, though *Blepharisma* cells do not ordinarily swim away from illumination. In nature they are found swimming between scraps of leaf litter at the bottom of a pond, and are often found free-swimming in the more brightly lit water above. The light-sensitive molecule is known as blepharismin, and if it is applied to non-light-sensitive cells of *Paramecium*, this ciliate microbe can acquire an ability to respond to light.

The light-sensitive pigments in *Stentor* and *Blepharisma* have been found in very different species. There is a similar molecule in St John's wort, *Hypericum*, and also in the buckwheat, *Fagopyrum*. Both are meadow plants. Farmers know that if cattle graze on St John's wort they can become so sensitised to light that, in extreme cases, sunlight can kill them. Here we have evidence of the same compounds existing in flowering plants and in single-celled animals. As protozoologists are conferring light-sensitivity on free-swimming cells which ordinarily lack this propensity, farmers are already observing the consequences of light-sensitisation in their cattle. Once again, we see common principles across the broadest spectrum of life.

Capturing an eye

The ciliated organisms discussed above are an intriguing bunch. They respond to light, and they know how much is good for them. Unlike the flagellates, in which microscopic eyes are widespread, the ciliates do not posses eye-spots. There is one interesting apparent exception to the rule. This is *Strombidium*, which is found in rock pools and lives a double life. At high tide, when its rock pool is flooded, *Strombidium* swims away from the light and burrows down towards the bottom of the pool. There it forms a cyst around itself for protection. At low tide, the process is reversed: the cell emerges from its cyst and becomes attracted towards light once again. In this form it swims upwards and carries on its rock-pool existence as a free-swimming ciliate.

Strombidium is a beautiful, complex little cell. Towards the front end is a series of complex organs for swimming and feeding, and inside the endoplasm near the tip of this extremity is a tiny, but unmistakable, eye-spot. How can one ciliate in all the microbe universe have an eye-spot, when the others are not so endowed? Patient observation reveals the secret. The mature *Strombidium* cells contain tiny green algae in a symbiotic association (as we have already seen, this form of co-existence is common in nature). In time, a senescent algal cell can be seen to degenerate until only the eye-spot remains floating free inside the host cell. Through some unknown mechanism, the eye-spot is moved towards the front of the *Strombidium* where it becomes part of the host cell's cellular equipment. Even if the ciliates lack eye-spots, at least one has found how to liberate one from a sighted cell and incorporate it into its own.

The most elaborate microbial eye

The dinoflagellates are amongst the most complex of all flagellate organisms. I have already referred to the way in which they sometimes form dense blooms in water. There can be millions of cells in a small glass of water, and many of them produce highly potent nerve poisons. There seems to be no specific purpose in this; it is probable that the nerve poisons are incidental by-products of the cell growing in crowded conditions. The occurrence of a chemical that reacts so strongly with mammalian nerve impulses has often been greeted with much surprise. A recurrent theme of this book is that there are more similarities than differences between plants and animals, large and small. In that sense, you would half expect components of an alga to interact with those of ourselves.

One of the most complex of all dinoflagellates is *Erythropsis*, a rare and highly fragile organism found deep in the oceans. It moves around not by beating its flagella, but by extruding a contractile arm (the *piston*) which it uses to explore the surroundings and crawl across the silt. Where *Erythropsis* exists, light levels are low, which makes it hard to find its prey. Its sense of sight is wonderfully developed. Its rounded ocellus is large, for such a cell, measuring about the thickness of a fine human hair. Within it can be seen all the principle features of a mammalian eye: a cornea, a lens, an iris to regulate light levels, and a retina. There is even a set of fine fibrils running away from the retinal cup, for all the world like a miniature optic nerve.

Consider this for a moment. Eyes in mammals are made of millions of separate cells, each with a specific task to perform, and all arranged in neat precision to form the eyeball. Specialised cells produce the transparent cornea, light-receiving cells (the rods and cones) collect light and colour information. This is fed to the brain by a nerve composed of thousands of separate nerve cells and their elongated extensions, where the information is processed and interpreted. *Erythropsis* is itself just one cell: it can have none of these specialised features. Yet look within it, and what do you see? An eye. It is looking back at you.

Navigating bacteria

Although most bacteria do not care for brightly lit surroundings, two rare groups contain pigments which enable them to harness sunlight, and these need a light-sensing system if they are to be in the right place at the right time. These primitive bacteria date back to the beginnings of life on earth. They are the halobacteria, and contain purple pigments with which they harness solar energy. They are closely related to organisms which produce methane. So primitive are these organisms, and so unlike other forms, that they are sometimes placed in a group of their own which is named to commemorate their supposed antiquity – the Archebacteria.

Halobacteria still exist in hot volcanic pools rich in minerals, and have flagella with which they can swim to ensure that they maintain optimal light levels. In a darkened medium they head towards the light, but if the light is too bright they change direction and swim away from it, until optimised levels are found once more. The lesson to be learned from these bacteria is that, for all their primitiveness, they embody some of the optimising attributes of more complex organisms. But here is the surprise. What is their photo-receptor? Some primitive form of chlorophyll, or a bacterial version of carotene which dates back to the dawn of time? The light-sensitive compound in the halobacteria is now known to be rhodopsin. It is identical to the compound in your own eyes, as you read this book. These most primitive of organisms use the same pigment to see as the most highly evolved creature known to science. Truly, the study of senses is a great leveller.

Some bacteria are now known to possess a sense in common with some vertebrate animals – they can detect the earth's magnetic field. These organisms produce tiny grains of ferric oxide Fe_3O_4 (also known as magnetite) which is a magnetic compound known to the ancients as lodestone. With the aid of these little grains they follow the earth's lines of magnetic force, and this enables them to swim downwards. The bacteria are known in both the northern and southern hemispheres, so it appears that they home in on the north magnetic pole (in the northern hemisphere) or the south magnetic pole (south of the equator). One question arises: which way do they swim at the equator, where the lines of magnetic force run parallel to the earth's surface? I am not certain anyone knows the answer to that, though one could facetiously observe that clearly (if they kept swimming long enough) they would end up in one or other hemisphere. None has been found in the equatorial regions; perhaps that's why.

The flagella which bacteria use to swim have been known about since the nineteenth century. Their structure is remarkable. Each is composed of a central core of two filaments with eleven more wrapped round the outside; the whole array is held inside a protein matrix which makes up the shaft of the flagellum. It is a little like a two-layered woven whip, and this structure is found wherever there are flagella. They operate by sending waves of movement along their length, much in the same way as a stage entertainer flips a microphone cable. Flagella, however, rotate about their axis and each ends in a rotor which lies within a stator inside the cell membrane, adjacent to the cell wall. This acts like a molecular motor, spinning the flagellum around.

The coordination of the flagella is precise, for they manage to avoid becoming entangled. The pattern of beating varies from time to time. The flagella with which *Escherichia coli* swims, for example, number about six on each bacterial cell. Normally the *E. coli* cell swims in a straight line. The flagella all beat together, each one spinning clockwise, and so well coordinated that they form a continuous bundle which propels the cell forward. In itself, this is a remarkable achievement for a cell so small. If the cell needs to change direction, the bundle is broken as the flagella separate and go into reverse thrust. In this mode they splay out from the cell, speedily arresting its forward motion and bringing it to a stop. The cell then changes direction, with the flagella beating anti-clockwise. The cell sometimes spins for a few seconds, tumbling in the water in an apparently random fashion as it senses which way to turn. Then, after a pause, the bacterium sets off again on a new heading. As this forward movement resumes, the flagella return to their coordinated beating as a bunch, and the cell moves forward gracefully as before.

Some of the secrets of this complex series of adjustments are becoming known. Methionine, for example, is known to be important for tumbling to occur. In a methionine-free culture medium, the *E. coli* cells are soon observed to cease normal tumbling altogether. The levels of calcium ions inside the *E. coli* cell are also important, and low concentrations of calcium are necessary in the fluid in which the cells swim for tumbling to occur. It might be tempting to assume that *E. coli* can sense which way to go by a change of concentration of a given substance on different parts of its surface, but that seems impossible. The cells are very tiny (only a few millionths of a metre in length) so that you could fit 50 of them across the thickness of a human hair. It now seems probable that they have a memory system which recalls the concentration of a molecule and compares it with the amount present as it swims along.

This makes it easier to understand how bacteria can function in their natural environment, but it reminds us of one key fact: not only can bacteria recognise desirable and dangerous compounds as they go about their affairs, but they have enough of a memory to know whether the concentration is rising or falling, and enough data-processing capability to know when to stop and turn round. We easily imagine that the tumbling response is a random affair, so that the cell takes up a trajectory determined by chance, not by intention. That may yet prove to be an over-simplification. An organism so well-equipped with sensory and response facilities may be making decisions that are more reasoned than we have suspected in the past.

How do microbes locate their prey?

There is a Chinese proverb which says: 'You may stand open-mouthed on a mountainside for many days before a duck flies in.' It is meant as a censure of those who are not sufficiently enterprising to get ahead with their plans, but it offers a lesson for all protozoologists. We too easily imagine that single-celled organisms move blindly on until they intercept their prey by chance, so that feeding becomes an accidental procedure. Life is more purposive than that. Studies of *Amoeba proteus* have thrown some light on how it identifies what it wants to eat. The experiments were done by feeding cultures of amoebae with progressively simpler food. At first they were offered the ciliate *Tetrahymena* on which they customarily feed. Later tests were done with cells of *Tetrahymena* from which the cilia had been removed, and which could not swim away. From there the trials moved to feeding the amoebae with single cells from other inhabitants of fresh water, including *Hydra* and freshwater mussels. The final tests were done with *Hydra* matter ground up and packed into microscopic plastic capsules with a porous surface, each of these microspheres a kind of 'artificial cell'.

No matter what food particles were offered, the amoebae consumed them. Empty microspheres were ignored. So were microspheres which had been boiled to denature the proteinaceous contents. Most interesting were experiments in which the *Hydra* material was packed into microspheres with progressively smaller pores. The smaller the pores, the less the contents could be detected; and the less detection, the lowered rate at which the amoebae responded. A further observation showed that amoebae never over-eat. The more abundant the supply of food offered to them, the less frequently they feed. Too much food acts as an inhibitor to feeding, quite the converse of what most humans get up to at parties.

It is clear that an amoeba can detect its food by sensory means, but it is also apparent that the feeding response is not automatic. Amoebae know how much they have eaten, and do not over-indulge. They sometimes make mistakes, however.

Occasional examples have been recorded of amoebae consuming grains of sand or ground glass. There are more subtle mechanisms that come into play. Tiny beads of a resin bearing positive electrical charges are readily ingested by the hapless amoeba, and so are microscopic sand particles if treated with positive ions. The amoeba is attracted to positively charged substances automatically, like a 'subconscious' response. However, it does not ingest these particles as though they were food. The amoeba simply glides over the particle and absorbs it into its cytoplasm as it passes over. There is no normal feeding ingestion. The effect is different.

Do not be tempted to imagine that these observations reduce feeding in amoebae to a simple, scientific matter of attraction. We have hardly begun to scratch the surface of a vast problem:

○ It may be true that an amoeba has a diphasic response to food – too little and it feeds, too much and it stops – but how does that two-way switch operate? What set of senses does the cell possess to allow it to tell the difference?

○ If the prey of *A. proteus* stimulates it to feed, why has the cell evolved to lose that ability? Is the attractant an inevitable by-product of normal metabolism?

○ *A. proteus* is a very slow-moving organism. Its prey, *Tetrahymena*, is a highly active, free-swimming ciliate. How does the gentle amoeba entrap something so lively?

This final point is an enduring mystery. All ciliates are active, and *Tetrahymena* particularly so. It darts about and spins through pond-water. Yet in a culture dish containing *Tetrahymena* and *A. proteus*, the amoebae always end up replete with ciliates. Some contain one, other amoebae contain several;

and this all happens within half an hour. It is an extraordinary performance.

The presence of the ciliates accelerates the rate at which the amoeboid cells move. This is clearly due to a sense of olfaction, because the addition of a culture liquid in which *Tetrahymena* are grown (but from which they have been removed) also causes the increased activity of the amoebae. If an amoeba senses a ciliate nearby, it extends a pseudopodium towards it. The end of the pseudopodium spreads out and forms a cup-shape which cautiously edges round the ciliate cell. The size of this cup is far larger than that of the *Tetrahymena* itself. The cup rapidly closes round the prey, and draws it towards the cell body of the amoeba. Sometimes the *Tetrahymena* struggles to escape, but *A. proteus* is used to catching these ciliates, and the amoeba is usually successful.

The sensing of nearby prey, the speed of coordinated response, and the effective entrapment of a dynamic and active ciliate by a slow-moving cell shows how much we have to understand about the amoeba. It is far from being a simple and basic organism, but one from which we have much to learn.

Sex in microbes

The popular concept of microbes reproducing by the simple expedient of dividing in two is only a small part of the story. Most species have a sexual phase. Ciliates ordinarily graze on bacteria and divide in two every few days. Once in a while, however, a sexual phase supervenes. The swimming cells pair off, their oral surface against each other, and fuse in sexual conjugation. They exchange nuclei – each individual fertilising the other, in effect – and later undergo multiple divisions. The resulting progeny then enter a conventional asexual phase of reproduction by division in two, as before.

What drives a ciliate to undergo sexual reproduction? The

cells are swimming and grazing, clearly a desirable state in which to exist, and are then diverted from this activity by an impulse to indulge in sex. Like humans, ciliates clearly derive more pleasure from sex than from feeding. We can argue about the notion of pleasure in a single cell, and you are at liberty to substitute other terms if you wish (satisfaction, metabolic equilibrium, instinctual gratification, whatever), but nothing can alter the fact that a single cell, when the urge is upon it, prefers mating to feeding.

The sexual imperative depends on the would-be mate in the microbial world, much as it does in human society. Cells select a partner, and sometimes try several possible sexual partners before union occurs. Protozoa can certainly tell one from another. If an adult *Vorticella* cell enters a sexual phase, it gives rise to male gametes by budding. They swim off, propelled by a circle of cilia, apparently inspecting other cells until they find one which can be identified as a suitable mate. Then (and only then) will they swim to the new cell, fuse with it and complete the sexual cycle. Unlike humans, ciliate sex often goes through a variety of stages. The most intricate life cycles pass through 20 stages or more, involving changes of shape and genetic constitution of bewildering complexity. Some have taken decades to untangle. In some cases, distinct organisms with their own specific names have turned out to be different stages in the life cycle of the same organism.

Even cell division is not as simple as it seems. There are several versions. Protozoa may increase in size and then divide into two identical daughter-cells (this is the process known as binary fission). Sometimes the reproduction is by budding, where the parent cell produces several small offshoots which take time to grow. There are also examples of repeated fission, where a cell nucleus divides several times, but the cell mass does not split until after the final division has occurred. Some protozoans form colonies, which never separate; others exist as isolated cells in some phases, and in colonial associations for long periods of time. There are complex sensory mechanisms which cause a cell to opt for one of these versions. We cannot yet determine what these mechanisms are, or how they function.

Paramecium provides an example of the complexities of sexual reproduction in ciliates. Two swimming mature cells begin to circle each other, each clearly weighing up whether the other is a suitable mate. If there is mutual approval, the cells approach as they swim and join together. As in many protozoa, there are two nuclei inside each cell. The macronucleus is large, and seems to be concerned with the daily running of the cellular machinery. There is also a smaller micronucleus, the true genetic centre of the organism. Only the micronucleus takes part in sexual congress.

As the two individuals join, the macronuclei in each cell break down and disappear, mingling with the cell contents. The micronuclei of each cell then divide several times – three or four divisions are normal, but the number varies with the species. All but one of the nuclei then disappear, and the remaining single nucleus splits in two. One of these divisions halves the total number of chromosomes (as happens in the formation of all normal sex cells, including the egg and sperm in humans). At this stage an exchange takes place. One of the two nuclei in each cell pass across into the other. The newly arrived nucleus in each cell fuses with the nucleus already present, and the fused nuclei then undergo several more divisions. After this, one of the nuclei becomes the new micronucleus, and the remainder fuse together to form the new macronucleus. Some are left over, and they dissolve away and disappear into the cytoplasm of the cell.

Sometimes an individual *Paramecium* will undergo a similar process without the intervention of any other cell. In this process, endomyxis, the macronucleus vanishes and the micronucleus undergoes several divisions. All but two nuclei disappear, and the cell then divides to form two daughter-cells with one nucleus each. These nuclei then undergo further repeated divisions, until new cells with a macronucleus and a single micronucleus are formed. The whole business is immensely complex and without parallel in multicellular animals. These processes are long and involved, all impelled by a series of unfathomable drives and sensory mechanisms within the *Paramecium* cells. They are triggered by molecules which are produced during the sexual phase of reproduction, and which are otherwise absent. The exact manner by which they are produced, and sensed by other cells, remains unconfirmed. Many of these sensory imperatives seem to have counterparts in the ways in which more complex animals interact.

Some insight has been gained from studies of the related ciliate *Blepharisma intermedium*. Chemical communication takes place between the two strains, mating types I and II. These are produced as indicators and are distributed over the cell surface. Mating type I secretes an indicator molecule, G-I, which reacts with other cells of mating type II and stimulates them to prepare for mating. The effect of G-I also alerts the mating type II cells to start secreting their own indicator molecule, G-II. G-II, when detected by the mating type I cells, prepares them for mating. G-II also seems to stimulate the type I cells to secrete yet more G-I.*

---

* G-I was characterised as a basic glycoprotein of molecular weight 20,000. G-II has been identified as calcium 3 (2' formyl-amino-5-hydroxy-benzoyl) lactate.

Algae that communicate

Some free-swimming types of algae live together in colonies. The classical example is *Volvox*, which exists in the form of spherical colonies easily visible to the unaided eye, and about the size of a bold full-stop. The colony consists of separate cells embedded in a gelatinous hollow sphere which they secrete themselves. Each cell sends out six fine fibrils, each of which connects to its neighbour. The effect is rather like an organic, living geodesic dome. The cells beat in unison, so that the globe turns sedately in the water and propels itself along. An encounter with an unwanted stimulus signals the cells to alter the direction of the beat, so the globe can roll off on a different course. These algal cells transmit nerve signals to one another to coordinate the movement. There is, however, no central authority which speaks for the rest.

Free-swimming algae are not the only forms which communicate. Subtle senses also exist in the algae that grow in the form of fine filaments. These are commonly known as 'witches' hair'. The well-known *Spirogyra* is an example. It customarily grows as rectangular cells joined end to end, a configuration which is perpetuated by the way each cell elongates, and then divides across the middle. *Spirogyra* maintains itself in this form for weeks at a time, producing filaments of great length. In time, sexual reproduction supervenes. Two filaments detect one another's presence through some set of carefully tuned senses. We do not know what they are, or where the sensors are found. One of the filaments is 'male', the other 'female'. Only they can tell which is which. From each filament, connecting tubes are produced from protrusions on the adjacent sides of the cells. The two tubes extend towards each other with the precision of tunnellers aiming to meet halfway. We do not know how they undertake this careful act of detection, nor how the alignment is obtained.

Eventually (and this happens along the filaments, as long as they are close enough to each other) the two protrusions meet, fuse at the extremity, and join to form an open connecting tube. The cells from the 'male' filament are squeezed through the tubes to meet the 'female' cells in the other filament, and the two cell bodies fuse. They are destined to secrete a protective capsule, and in this form they will survive the winter and germinate, producing a new filament in the spring. The method by which the migrating cell is propelled through the conjugation tube has been known for many years. Tiny fluid-filled vacuoles form along the side of the cell facing away from the intended direction of movement. These pump water into the space between the cell contents and the cellulose cell wall. As the pressure builds up, the soft cell is forced down the tube and through to the other side. It is a matter of applied hydraulic pressure.

But what of the senses that lead to this process? How do the filaments tell where they are, and what directs the growth of the conjugation tubes towards each other with such extraordinary precision? Above all, and perhaps of the greatest interest, where is the seat of the sexual identifier – the sense that can inform the filaments which is 'female' and which 'male'? There are many factors involved in this complex mechanism, and some of them indicate the existence of highly attuned senses unknown to humans. Algae *can* communicate.

Ciliates

There is one group of the protozoa which deserve special attention. These are the ciliates, which use a surface coating of thrashing microscopic hairs; sometimes to catch food, sometimes to swim. Ciliates are remarkable objects for microscopic study. How do they function as they do? A single swimming cell, if it encounters a solid object in its path, reverses for a short distance, turning as it does so, and then sets off on a different track. It looks like any other animal attempting to avoid an unseen obstacle. Familiar animals, though, have senses we can recognise, while the single-celled protists remain to be properly understood. Watching these organisms under the microscope makes it easy to forget that they are single cells, for their movements seem more purposeful by far.

The first attempts to analyse the causes of the movement were based on the methods used to examine mammalian muscle. Cells of *Paramecium* were chilled and soaked in a solution which removed their ions and left them intact, but effectively dead. In this way all the functioning components in the cell were left, but there were no free chemical components to fuel them. It was a simple matter to infuse the cells with substances applied from outside; in this way, the experimenters could decide which ions were present within the cell and not the *Paramecium*. The cells were treated with very low concentrations of calcium in the presence of magnesium and ATP (adenosine triphosphate, the cell's principal source of metabolic energy). At very low calcium concentrations the cilia on the cells started to beat strongly, driving them forwards. As the amount of calcium was slowly increased, the cilia began to alter their direction of beat until the cells were being driven backwards. If no magnesium was present, the cilia would move to take up their position ready to beat, but no further movement was seen. In this way we can deduce that calcium ions are vitally important in deciding the direction of the beat, but magnesium is vital for the beat to take place at all. In further research, microscopic electrodes were inserted into *Paramecium* cells. The nature of the electrical signals generated within the cell depended on where the cell was stimulated by touch. Contact with the front of the cell was followed by a wave of depolarisation of the cell membrane – in effect, it momentarily lost its electrical charge. If the rear of the cell was touched, the opposite reaction occurred, for the cell membrane increased its polarisation.

From observations like these we can sketch in what goes on within a *Paramecium* cell. In normal swimming it moves forward gracefully through the water until it encounters an obstacle in its way. This causes the momentary depolarisation of the cell membrane. Without an electrical charge to keep calcium ions at bay (see p. 240), extra calcium can migrate inwards to the cell, raising the concentration to a level high enough to trigger the reversal of the ciliary beat, and the *Paramecium* moves backwards. The polarisation is transitory, and in a short while the normal conditions are restored and excess calcium is pumped out again. Normal ciliary action is thus restored and the cell carries on swimming forwards (but in a different direction from last time). If the cell is touched from behind, the raised polarisation of the cell membrane drives out even more calcium, which has the effect of intensifying the forward beat. The *Paramecium* speeds up, and quickly swims away from the source of the stimulus. We can thus model the control of movement in these ciliated cells. The polarisation of the cell membrane, and the role of calcium, is as important to *Paramecium* as it is elsewhere in the world of animals.

Humans revert to the microbe state at reproduction, for an egg and a sperm are single-celled organisms. The sperm acts like a flagellated microbe, and we are just discovering that a wave of calcium ions passes through the egg cell at the moment of fertilisation. If calcium is released into the unfertilised egg of a sea urchin in the laboratory, it starts to divide as though it had been fertilised by a sperm. It has also been reported that calcium serves to set a human egg dividing after fertilisation, for the sperm brings calcium-releasing agents as well as its genes. Perhaps a lack of calcium response explains some cases of infertility. It may be that artificially controlled calcium surges could start an unfertilised egg-cell dividing. We already know how important calcium is for plants. We have seen that it is important in the control of cell chemistry and it is no surprise to learn that this is true of all cells, plant or animal, specialised or not.

However, just because we can see how movement is controlled, this does not mean that we should too hastily conclude that the resulting movement is a merely mechanical response. Ionic movements are equally important in the regulation of human musculature, yet we still coordinate these responses to take a book from the shelf, turn the pages and settle down to read. Protozoa can be seen nuzzling through the detritus at the bottom of a pond, searching out the items they wish to consume and drawing back from unwanted encounters. There is nothing mechanical about that. These protozoa are thinking, as far as they have to, and processing information. We stand to learn much about the individual cells of which we are composed by a study of the ways of protozoa.

An enduring mystery about the protozoa (and particularly the ciliates) is where they come from. One of the most popular means of starting a culture of ciliates is to boil some hay and leave the resulting brew to stand. Within a week, the water will be swarming with protozoa. This indicates that they can form resting spores or cysts which can withstand heat, and which hatch out when the water cools. Obvious as that may seem, cysts are unknown for most ciliates. They must be among the commonest of phenomena in these organisms, and yet have rarely been documented and for the majority of types they are completely unknown. Some ciliates certainly come close. The ciliate *Spirostomum* is more than a millimetre long and as thin as a hair. But where do they go in the winter? I studied a colony to investigate how they vanish in autumn, and reappear each spring. If *Spirostomum* is slowly cooled in the laboratory as though winter were approaching, it sinks to the bottom and communicates with its fellows to collect together. These were later found among the leaf litter at the bottom of the pond, where they gather together in large groups the size of a soft peanut. Within this mass, the cells pull themselves in and begin to round off. The end result of this process is that the *Spirostomum* cells lose their ability to move and enter a resting phase in which they are perfectly spherical.

When I warmed them again, they began to lengthen and eventually to resume the active swimming phase. Tantalisingly, we could not induce them to form actual cysts, with a resistant protective capsule. For the present, the fact that they form a resting stage at all is interesting enough. However, a resistant resting stage of ciliates must exist; so common are they in infusions of straw that they were originally named the Infusoria. It remains odd that such a strongly inferred phase of the life cycle is never normally witnessed by the human observer.

It is always possible that we shall discover cysts if we start to look for them. The Victorian microscopists were the greatest students of the ciliates, and made detailed studies of them and their life cycles that have rarely been equalled. In the twentieth century we have preferred to find mechanistic models to explain what we observe: changes of pH (acidity), a raising or lowering of dissolved oxygen, the movement of calcium ions across a membrane. Many of these phenomena also explain how humans work in adapting to light, moving a muscle, responding to food or changing the way we breathe. However, for all the simplicity of the underlying principles, we retain our special qualities and would not regard ourselves as mere automata.

Mechanical modelling in our studies of single cells can go too far. We understand much of the way cells undertake the basic processes of life, but we know little of how those are coordinated into the great web of living. Protozoa look for all the world like multicellular animals as they move around and explore their world, and there is no mistaking the sense of purpose in the way they behave.

Seat of movement

Our own movements take place through our muscles, and muscles function through the action of actin and myosin, proteins which have the power to contract. These two components combine to form a complex, usually known as actomyosin, which makes muscles move. Protozoa have fine threads of actin and myosin inside their cells, and this is what drives the beating cilia and the lashing flagella that move the cell through its medium. The actomyosin complex is found throughout the animal world, and proves to be the source of movement even in amoebae. There was a belief that amoebae moved by the regular liquefaction of cytoplasm as the cell moved forwards, followed by the formation of a firmer gel in the cell body. Now we know better. Fine threads of actomyosin exist throughout amoeboid cells, and their contraction causes the cell to move. The threads are not permanent, but are produced by the amoeboid cell when it needs to move.

Actomyosin acts in a simple manner. It splits ATP and metabolic energy is immediately released. The energy is used by the actomyosin to contract, and this is the secret of the way muscles work. Much research has been done with the slime mould *Physarum*, and one experiment showed how similar actomyosin from plant and animal sources truly is. Actin from the microbe, combined with myosin from animal sources, gave a form of actomyosin which functions perfectly normally.

Where else might we find actomyosin? Within plant cells, active streaming of the cytoplasm takes place. The stonewort

Chara contains giant cells, and there are also larger-than-normal cells in the hairs which pack out the stamens of *Tradescantia virginiana*. Analysis of these cells also shows that the movement is caused by the coordinated application of actomyosin within the cells. There are general principles behind these observations. All living things move. Microbes, whether plant or animal, are often in active movement from place to place. Those that stay rooted to the spot exhibit movements within their cells. In every organism the cause of the movement is the same compounds that are present in our own muscles. We are bound to different forms of life through common properties. This universal principle of movement is often regulated by outside forces. External stimuli are recognised by acute senses, and the responses are normally appropriate to the nature of the stimulus. It is true that we can model movement through the liberation of energy from ATP, or the pumping of calcium ions across a membrane. However, this merely explains the chemical origin of the movement and says nothing of the complex network of responses that regulates it.

The amoeba develops tracts of actomyosin in order to move, and has mechanisms to know exactly how to do that. These cells sense their surroundings, and adapt their movements and their lives as a consequence. Humans have a unique set of mental abilities which set us apart from other organisms, true. However, our senses and our innate abilities are much the same as those of the supposedly humble microbe. The remarkable thing is that an amoeba can do most of the things we ourselves can do. It can also carry out a number of vital tasks which are quite beyond us. We would do well to look at all forms of life as serving a united purpose, and having a common aim. Senses are ubiquitous, and are as important to a protozoan scurrying about in its pond as they are to the reader scanning this book.

Complex senses in simple animals

We have seen how the myxamoebae come together to form a single animal, and how algal cells can produce the colonial structure of *Volvox* and its allies. Cells have learned how to cooperate. As they discovered the benefits of living in communities, specific cells were given specific jobs to do, and became specialised for a purpose. The Parazoa are the sponges. They contain cells which use their flagella in a coordinated fashion to waft a supply of water through the body of the whole sponge. This gives them fresh oxygen, carries away wastes (dissolved in the water) and provides the cells with food particles. If a living sponge is broken down into its component cells, they can be sieved through a delicate mesh big enough only for single cells to pass. Once in their new environment, the separated cells seek each other out again and rebuild a sponge, much as existed before.

We can only speculate on the complex interplay of senses each cell possesses. They need refined senses to recognise each other and know how to cooperate in rebuilding the entire colony. Cells have a great capacity for adaptation and specialisation, and this is clearly in line with stimuli received from outside. The sponge cells' capacity for reorganisation reminds us that every responsive cell must have a full set of appropriate senses, and a means of processing the data to evince an appropriate response. The cells around the pore of a sponge through which water is admitted become muscular, and if an unwanted stimulus is detected they contract in an instant, closing the pore and preventing some noxious substance from entering the sponge's internal spaces.

There are other little-known organisms, including the cnidaria and ctenophora – small, jelly-like creatures which show how life first began to evolve into multicellular forms. The senses are highly developed in them all. The cnidaria are among the many types of animal that possess nematocysts, remarkable cells that can turn inside out and project, at high speed, a pointed harpoon into their prey. It has been said that nematocysts fire automatically when touched, but the reality is more complex. It is now clear that their activity is related to the physiological state of the animal and its level of nervous stimulation. Some species have fish that live amongst their tentacles. This is also observed in sea-anemones. The fish are not attacked by the nematocysts, which fire at other fish but not the commensal species. This suggests that even 'lowly' polyps and their relatives can sense one kind of fish and adapt their behaviour accordingly.

In the flatworms, the sense organs congregate towards the front of the animal. This is the process of cephalisation, or head-formation. Flatworms show complex abilities to respond to the outside world. They recognise each other and mate with a chosen partner. If they are cut up, each portion will grow into an entire new flatworm. Yet they can learn, and it is possible to train them to respond to recognisable stimuli. Flatworms have a primitive central nervous system, and show many of the traits visible in vertebrates. Clearly, they possess a range of senses, and some sense of what to do with the stimuli they receive. Many species of marine animal still have ciliated embryos which look much like protozoa. This morphology embodies resonances of that earlier stage of animal development. We know that ciliates distinguish front from rear; these embryos can clearly tell upper from lower. As they grow, some of them enter a phase where they become crawling organisms existing on the surface of the mud in a pond. Before this they undergo some flattening, and always swim the right way up during this period of change. They must be able to orient themselves through the use of their senses.

An important attribute of some animals is known as thigmotaxis – the tendency for the creature to keep the largest possible area in contact with the substrate on which it lives, like a limpet holding on to a rock. This is a process with clear benefits to the organism, particularly in tidal areas or regions of streams and rivers where there are strong currents, but little is known of how these creatures select a suitable surface or maintain their hold. Limpets, for example, have paired eyes and although they browse around the rocks grazing on algae when the tide is in, they always return to exactly the same spot as the water begins to recede. Not only does this say much for their perceptual sense, but they clearly have a sense of timing too, in order to maintain the maximum grazing period while still having enough time to return to home base in safety. This is a complex set of activities, and, barring accidents, they always manage to complete the process.

Some of the semi-transparent little worms have photo-receptors in the brain itself, while others have them on the tentacles which they use to explore their world. These little arrow-worms which live in the sea are fearsome predators which float amongst the plankton layers and hunt their prey from among the little larvae which abound. These creatures are transparent, free-swimming marine animals of the phylum Chaetognatha. They are a few centimetres in length and streamlined, and propel themselves along by quick darting movements produced by paired fins down the sides of the body. There is a complex system of bristles around the head, with which food is captured, and a unique hood which covers the head when the arrow-worm is moving quickly, to help streamline its contour.

Arrow-worms form huge communities around the coasts and across the continental shelves, and are an important food source for larger animals. For all their simplicity, they have an excellent sense of movement-detection, for they can accurately locate creatures around them from vibrations in the water. An arrow-worm finding suitable prey within reach turns its body to face the target, and then strikes with great speed and seizes the creature in its jaws. They are attracted to organisms which emit vibrations in the range of 10–20 Hz. Arrow-worms can attack creatures larger than themselves (young fish, for instance). The senses of these eyeless creatures are clearly able to distinguish the correct size of prey. If an organism is too small, the arrow-worm ignores it, but if it is big enough to pose a threat, the arrow-worm speeds off in the opposite direction.

Proprioreceptors are organs which enable an organism to detect its position, relative to outside reference points. It is a convenient term, but the complexity of such a sense is hard to explain in scientific terms. It clearly works: there are specific depths at which you might find an organism, or a particular kind of substrate. That is, indeed, where you find them, all pointing in exactly the direction you would expect. They all catch their food on cue, too, though we do not know exactly how that cue might operate.

Jellyfish manage to maintain their position in exactly the kind of environment favoured by their food, and they do this by swimming purposefully when the need arises. Some of them have tiny ocelli, though others are blind. There is said to be a 'dermal light sense' which allows organisms to pick up light on the skin surface, even without any identifiable photo-receptor organs. It may explain how organisms maintain an optimum distance from the brightest regions of the water, for (though many are attracted by light) they can avoid light that is too bright.

Many primitive creatures can sense vibrations, which may help them to find one another through the beating of cilia in the water. Some use this sense to maintain their distance from the waves breaking on the shore, while others can detect prey through the vibrations they transmit as they swim. We must conclude that these organisms can interpret the movements which they pick up in the water. The movement of prey nearby makes them swim towards it; yet the vibrations of water striking the sea-shore make them swim away to minimise the risk of being stranded. Not only do these organisms listen out for vibrations in the water, but they know what it is they are hearing.

There are many other factors that influence the behaviour of these floating animals. Light and vibrations in the water are two of the most important, but they can also detect oxygen levels and pressure. They can taste the water, too, and pick up the chemical products of nearby organisms. This helps them to scent their prey, and to find their way to one another. In the types that live in fresh-water habitats, an ability to sense the acidity of the water (pH) is also important. Some of them show a preference for moving water; some species of flatworms are always found where the water is still, but others prefer to live in the gravel of a fast-moving stream. If their senses were simple, and their responses largely mechanical, we would expect them to be swept by currents into areas of relative stillness. That is not the case: they sense the speed of the passing water and keep in the fast-flowing areas they prefer.

There is also the question of sensing electrical and magnetic fields. Some of the organisms that live in the film of water surrounding soil particles seem to detect the electrical behaviour of root hairs, and use these fields to home in on roots they can parasitise. Some experiments have even suggested that flatworms can detect a magnet, and seem to be sensitive to the earth's magnetic field. We have seen that some bacteria are able to do this, and there is no immediate reason to doubt these suggestions.

Do you wake up before the alarm goes? You are not alone. Even these organisms have an internal biological clock. They can respond to the cycles of day and night. They are sensitive to the phases of the moon, and many of them rise to mate when signalled by a full moon at the appropriate season. Cycles of winter and spring, summer and autumn are detected by many organisms. Some sand-dwelling worms swim to the surface in such numbers that great areas of the sand become slimy with their presence. If they are disturbed, they retreat immediately into the sand, and move to deeper layers in proportion to the strength of the stimulus.

One sensory structure comparable to those in vertebrates is a cell containing tiny granules which monitor the orientation of these creatures. These cells are the statocysts, and they are rather like the middle-ear arrangement in ourselves. Combined with the sense of time, these help an animal to rise and fall in the water according to its needs. If animals with a daily or tidal rhythm are removed to an experimental tank where conditions are constant, the behaviour keeps up for prolonged periods, providing further evidence of a biological clock. During the hours of daylight, when strong sunlight may be damaging, many marine organisms sink to a lower level in the water where they live. As the sun sets they float higher, and come into contact with the plankton living in the uppermost layers. Perhaps the 'dermal light sense' operates in many of these species too, for not all of them are known to have photo-receptors. If that is the case, there may be other senses of which we remain unaware. As is often the case, different creatures respond in different ways to the same stimulus. Bright light inhibits the sexual reproductive phase of sponges, for instance, while it triggers the same phase in some little polyps. The length of daylight matters to some creatures. If some rotifers are exposed to long periods of daylight they tend to switch reproductive behaviour in favour of producing long-lasting eggs of the kind that allow the species to over-winter. Short lengths of daylight per 24 hours switches them away from this pattern and back towards the normal method of reproduction without sex (rotifers are one of the many types in which parthenogenesis, 'virgin birth', is common).

Communities

Senses are vital in the regulation of communities, and many less-developed organisms rely on a community structure for their survival. Flatworms, like nematodes, can collect together in groups, a phenomenon sometimes aptly described as swarming. We have seen how ciliates gather together in colonies as winter approaches. In some organisms the release of chemical pheromones has been documented. These are the chemical cues emitted by mammals as mating signals.

Which signals are used in detecting members of the same species and distinguishing them from others, or from prey? How are those signals detected? Once communities form, what effect does the preponderance of similar organisms have on the senses of each one? How does one organism respond to the presence of an unrelated form of life? We see examples of this in associations between hermit crabs, the shells in which they live and the sea-anemones sometimes found attached to the shells.

The sea-anemone *Hydractinia echinata* is usually found attached to the whelk shells that hermit crabs prefer. When first observed, you might assume it is a quaint example of erroneous attachment of a polyp onto a substrate which happens to move, rather than a stationary rock surface. That is not the case, for the species is adapted to grow on whelk shells containing hermit crabs. In some associations between an anemone and a crab, the crab is seen to detach the anemone and hold it in place with its pincers until the anemone is firmly attached. A slow discharge of electrical signals is used by other crabs to communicate with the anemone and encourage it to transfer to the shell. In other cases it is believed that the anemone is attracted by chemical stimuli resulting from the crab's effect on the whelk shell. There is some mutualism in all this, including the combined interest in capturing food and the benefits of transport for the anemone, whose stinging tentacles may help to protect the crab.

There are many special senses which these creatures possess. Some sea-anemones have symbiotic algae living amongst their cells. The algae can control the behaviour of the sea-anemone. Individuals free of the algae expand and contract at will, unrelated to the effects of daylight and more closely geared to the environmental conditions. Once the anemone has been colonised by the algae, this pattern abruptly changes. The anemones colonised by algal partners close up in the dark and open wide in the light; however, if the sunlight becomes too bright, they close up again in a typical avoiding reaction. These anemones have potential responses they do not normally use, but they can sense stimuli from their algal symbionts and can then react in a different way which benefits the alga, rather than the anemone itself.

Other communal pairings take place when one organism has designs on another, and this is the principle of the parasite. Although we do not lay much emphasis on the fact, parasites must have sense organs and highly attuned senses if they are to bridge the gap between one host and the next. Some parasites pass through a series of different stages in their life cycle. How these stages evolved is a major puzzle for biology, but how the parasites sense the need to make the move to the next host in the sequence is a rarely raised matter which justifies attention. In some cases (like an egg laid on a food-plant) the passage from one host to another is simple to understand. But in many cases the host has to be identified, and this demands acutely accurate senses.

Some of these life cycles are immensely complex. One parasite enters a sturgeon through the intestine, and later finds its way to the ovaries. Another has a complex cycle involving (at different stages) a snail, a seagull and a herring. The parasite has a lot to do. It has to identify the conditions that make transfer to an animal host most propitious. It then has to recognise the host. Once ingested, it responds to the circumstances under which it should emerge and start its new life. Many parasites have complex pathways they have to follow inside the host – colonising the liver, for example, or rising from the lung and being swallowed. The parasite has to find its way around the host's body, in cases like these, which is a feat of navigation calling upon a great deal of sensory sensitivity. The parasite may have to tell a vein from an artery, and know when it has encountered a liver cell, rather than one in the pancreas. I know students who have difficulty with distinctions like those.

A parasite that lives inside its host is often little more than a reproduction machine. It needs no defensive systems against marauders, or camouflage to avoid predation; it needs little in the way of musculature, no eyes or ears; and it doesn't have to find food, for it is surrounded by a never-ending supply. As they lose their faculties these parasites become increasingly simple, an example of evolution in reverse. None the less, the sense organs of parasites are amazingly varied. Those that hang on by means of a sucker or an adhesive gland have sense organs that can monitor the surface onto which they are holding fast, and check it. They can detect the degree of adhesion, and make adjustments. Liver flukes have special organs of taste around the sucker, to seek out the best place for attachment, and arrays of nerve endings stacked up inside little sense organs, papillae, have been seen in other parasites.

These papillae are well-developed in parasitic worms and flukes which need to meet to mate. Not only can they detect their own kind in the turmoil of the host's body cavities, but they can align and copulate through the use of their senses. They also know where to fix themselves. One worm attaches itself initially in the middle of the intestine, but in later life moves higher up, towards the host's stomach. It may be detecting bile, for in infections in hosts where the bile duct has been closed, it moves down the intestine and takes up a position nearer the anus. Within the gut is a swirling environment of changing pH and chemical composition, and parasites must not only resist being digested (or attacked by an immune reaction by the host) but must also know where they are, and where they need to go.

Some species of the parasitic worm *Schistosoma* infect humans, others infect rats. The human varieties are released into water during the middle of the day, when humans are most likely to be bathing and to acquire the infestation. Rats, by contrast, emerge in the early evening, which is when their parasites are released. Blood parasites of humans which are spread to new hosts by mosquitoes that fly during the day-time flood the bloodstream when the mosquitoes are most abundant. Species which infect monkeys (and are spread by insects which come out at night) predominate in the bloodstream during the hours of darkness. These tiny parasites are able to determine the time of day from within the tissues of the host, and adapt their behaviour accordingly.

In decrying the application of human models to the animal world, we can run a real risk of turning our backs on acknowledging their nature and see them, instead, as simple chemical systems which respond like automata to the outside world. Humans have their own refined senses, yet so do the smallest of creatures. Our anthropocentric era encourages us to worship humans as though we were not uniquely human, but uniquely alive. Look again at the lives led by these tiny moving microbes, and celebrate the ubiquity of the senses. They are to be marvelled at, not consigned to a wilderness of human-centred ignorance. There are many implications for the realisation that non-human animals are highly sentient. All animals and plants sense their surroundings, and thus they all have feelings. As we have seen, we have little understanding of the senses of many of them. How much sentience do they show? The answer is crucial to our future understanding of the world: all species have just as much intelligence as they need.

Only one species, in my view, has more than it knows how to handle.

7

A New Understanding for a New Millennium

With a new sense of humanitarianism we can rekindle the study of all living things. Schoolchildren should recognise the plants that grow in their surroundings; everyone could reclaim a knowledge of nature which has always been the mark of a mind at one with the environment. The last few centuries have brought us glorious revelations, from which modern biology too often retreats. The eighteenth century saw the birth of the Industrial Revolution in Britain and the dawn of physiology. In the nineteenth we saw the revelations of electricity, the birth of bacteriology and the cell theory. The twentieth gave us the atomic and electronic eras, running alongside gene mapping and the electron microscopy of the cell. This should have been an era of the flowering of the study of life, but instead it has become so narrow-minded that some people even seek to replace the word 'biologist' with the more high-tech term 'bio-scientist', a perfect indicator of the way we have begun to lose contact with the breadth of the subject.

The effects of this attitude have pervaded our culture. One example may be found in *Powers of Ten*, a marvellous swoop through reality from the cosmos to the atom by Philip and

Phylis Morrison of *Scientific American*. It is a challenging book, and a classic which deserves its place in the literature of science. As it explores the human body it takes views through the microscope of increasing magnification, a stunningly vivid way of presenting the secrets of human life. Crucially, the pictures they obtained for their book miss out the most revealing points of all. Under successive magnifications, the surface of the skin is revealed in beautiful detail, like random paving stones; a further increase of magnification then shows a single cell beneath the skin.

However, all we see is the exterior of the cell. It is a view through an electron microscope of a dead, metal-coated lymphocyte. There is no sign whatever of the vibrant translucence that marks out a real living cell. A further increase in magnification shows, once more, the appearance of a dead chromosome; after that is a computer model of the genetic code of DNA. The secrets of life are to be found within the cell, yet all the reader sees are blank masks of the surface. There is no sign of life in this part of the book; everything is a replica, and it is all dead. It is as though you were to produce a book on the working of hospitals, where a photograph of the exterior masonry was followed by a picture of the tiles in the bathroom – but which omitted the bustle of life, the patients and nurses, the drama of the operating theatre and the complexity of administration.

Domestication of animals

Humans possess a unique intellect, and an unmatched ability to relate to other forms of life. The relationship between ourselves and our domesticated animals is one of mutual benefit. In the earliest times humans travelled on foot, while wild horses were prey to carnivores and obliged to live out their lives in the face of extremes of wind and weather. Four thousand years ago the peoples of Asia discovered how to ride, and the domestication of the horse soon followed. From that time on, the mutual benefit was clear to see: humans were able to gain height, speed and relative safety, while the horse gained care, a more certain food supply, and relative immunity from predation by wild animals. The majesty of the horse conferred supernatural powers on them, in the minds of many peoples. Horses were often regarded as deities, and the centaur – half horse, half man – may have derived from the sight of mounted riders by tribal peoples unfamiliar with the domesticated horse. The bond between the horse and its rider is a unique synergy. It can certainly reveal something of the unspoken language which both species learn to share.

There is a greater question I would like to pose: it is accepted that domesticated animals become accustomed to human ways through classical Darwinism – the survival of those fittest to adapt to the ways of our society. I doubt that this can be the case. The differences between, say, a wild species and the domesticated dog are considerable. To believe that man has simply selected animals because of their temperament is hard to reconcile with the relatively short time that the two species have cohabited. It also implies that wild creatures must covertly embody the behavioural patterns that fit them for domesticity, an unlikely evolutionary trait. In many ways it is more feasible that domesticated dogs acquired their behavioural characteristics and passed them on to their offspring. This is the theory of the inheritance of acquired characteristics, the pre-Darwinian view of Jean-Baptiste Pierre Antoine de Monet, Chevalier de Lamarck (1744–1829). The theory was superseded by natural selection, and science accepts that new characteristics appear because of mutation and selection, never because of acquisition. The pioneering findings of the early geneticists, starting with the cross-breeding experiments with peas carried out by the

Moravian monk Gregor Mendel (1822–84), led to the view that genes pass inherited characteristics to each succeeding generation. In that model, acquired characteristics cannot be transmitted through the genes. If that is the case, then the behaviour of dogs could not be transmitted via DNA.

Yet there is a possible mechanism by which acquired characteristics could perhaps be passed to a succeeding generation. There are genes within the mitochondria which pass directly from mother to child. The mitochondria lie in the cytoplasm of the egg-cell, the ovum. The cell mass of the ovum expands during maturation, using components that are contemporary with experiential input for the mother. In this way, there is clearly a possibility that some modification during the life of an organism can be transduced into a form capable of transmission to a subsequent generation.

There is certainly much worthy of study in the various breeds of dog. Commercial breeders will accurately assign behavioural traits to the various types – pit-bull terriers are savage, labradors benign, and so on – and we do not stop to consider what this implies. First, it suggests that there is a pronounced and reliable hereditable component to character. There are many potential implications here for the widespread popularity among infertile couples of *in vitro* fertilisation by the use of donor sperm. Second, it is intriguing to relate the behavioural traits of a breed of dog to the dog's morphology and deportment. Savage breeds look savage; benevolent animals tend to have the deportment we recognise as unthreatening. Is this because the genes mediating behaviour are inextricably linked with those that confer external appearance, or is it because both are acquired by generations of domestication, and breeders link the two? It may be that the inheritance of acquired characteristics is a biological possibility. If it is, then it would be the behaviour of sentient animals that provided the clues.

Value of life

There are plenty of people who already celebrate the flourishing varieties of life, of course. Gardeners learn to understand their plants and to respond to their needs communicated through subtle signals. Animal-lovers have an emotional intimacy with their pets which can sometimes transcend faith in their fellow-humans. At a meeting on animals I once had a comment from an eminent scientist in the audience which showed how often people forget the strength of the relationship between animals and pet-lovers. 'There is one easy test of how we feel about animals,' he announced. 'Imagine you drive round a corner at night in a narrow lane, and have to choose between hitting a person or an animal blocking the way. That's when you realise where your priorities lie.' I tried out this question in a rural pub that evening. A third of the locals weren't immediately sure which they would choose. Two or three said they would prefer to know the identity of the two potential targets, and one insisted that humans were always worse than animals, for humans committed acts of which animals were innocent. She would risk demolishing a person, rather than an animal, any day. To individuals uninterested in pets, and intolerant of animals, such an attitude seems bizarre, yet there are many who prefer animals to humans. We tend to be dismissive of views to which we do not subscribe, but the devotion of such people to animals is a real phenomenon. For a lot of people, when family and friends are few, the psychological support of their pets can be of vital importance. It is easy to dismiss those who say they find animals more attractive than humans, but that may be more a reflection on the people they know than on any inner peculiarity.

Forbidding any harm to all animals has even become a tenet of religious belief. This seems to fly in the face of nature, where harm by one species to another is an essential component of survival. I have seen religious booklets which show predatory animals in a cosy relationship with farm animals and children, a modern icon of the vain promises of the Book of Isaiah that 'the wolf also shall dwell with the lamb, and the leopard shall lie down with the kid, and the calf and the young lion . . .' In nature, sentient animals are hunted and killed, rejected and eaten, stalked and dismembered by their prey. Killer whales toss broken-bodied sea-lions around as playthings, just as cats bring in dying mice to play with in the living-room.

In any event, a boundless love of animals tends to find its limits in the form of rats in the attic or woodworm in a wardrobe. Ardent lovers of plants rarely extend their benevolence to weeds in the gravel drive, or to dry rot fruiting in the cellar. It is the use of animals in scientific experiments that causes the most violent protests. It is hard to find any aspect of modern medicine that is not indebted to animal experimentation. Here too, as in farming, it is not enough for us to dismiss animals as though their feelings were of no consequence. Those who hold to the incalculable superiority of human life must accept, as a concomitant of their belief, that our moral senses are similarly elevated. A civilised person is also humane, and one way in which we express our humanity is through the avoidance of wanton cruelty to all living beings. Although animal experimentation is closely regulated by statute, there is a modern move away from the use of whole creatures in favour of tissue cultures or other *in vitro* modelling systems.

Yet there are also experimental procedures which a humane attitude to animal life would dismiss as barbaric. A major toxicity measure is the LD-50 test, used for many decades to assess new products. The acronym is an abbreviation of 'lethal dose 50 per cent'. The test is done by administering varying levels of a new compound to animals (often albino rats) and noting the

302   The Secret Language of Life

dose at which 50 per cent of the experimental animals die. The LD-50 result is a direct indication of the toxicity of the compound to the creature. There are aspects which make the test highly unsatisfactory. One is that the test measures toxicity to rats, which is not necessarily an indication of its poisonous effects on humans. The differing effects of drugs on disparate species led toxicologists to conclude that Thalidomide was safe for pregnant women to take, for example. In human patients it produced a failure of limb development, and its terrible consequences are with us to this day.

Humane observers condemn the test for its essential cruelty. If half the animals die at the given dosage, it follows that a considerable proportion have almost died, and will have suffered in consequence. The principal justification for this procedure is that rats are 'only animals', so the consequences hardly matter. That is not good enough. Our duty is surely to nurture the interrelated webs of life on our planet, and to employ a proper duty of care for the animals in our charge. These animals are sentient, and we should not condone cruelty to them. It is interesting that rats are so often used. We have an instinctive dislike of rats (perhaps because they do not seem to be readily intimidated by humans) and I have no doubt that an LD-50 would not have become so widely accepted if kittens were used, or chimpanzees. A humane era will end such barbaric practices.

A resurgence of biology, and the celebration of life in all its glorious variety, will revive our interest in the plants and animals that surround us. The greatest need is for us to understand life, to realise that we must interact with living organisms, and not always seek to dominate them. There has been a tendency to exterminate life when it seems beyond us. In America there was a campaign for the universal use of insecticides, so that our 'insect foes' might be completely eliminated. Yet we need insects: they are, for example, one of the principal agents of pollination. In Europe there was a move to make rivers into 'clean-water conduits' by banishing all life from river-water. What we have not understood is that it is the microbes in the water-courses that purify them and make them clear. If we had set about sterilising rivers, then the first algal cell that alighted in the water would have triggered a suffocating bloom of pea-soup consistency.

Our aim as humans must be to value the global network of all plants and animals, and react to their presence with respect. The unravelling of a new millennium offers a chance to espouse a new vitalism, and a revolution in life sciences: the reduction-ist sciences of molecular biology and genetics are not enough, for what we need is a full understanding of how living organisms interact. The nineteenth century gave us the first full flowering of microbiology, when the extent of microscopic life began to emerge. During the twentieth we have embarked on a mad scramble to publish papers on single, small effects. Now is the time to take stock and create scientific understanding which fits the data together. For the twenty-first century we need a new holistic view that unites discrete findings into new patterns of understanding.

It is true that people sometimes misunderstand animal wel-fare. A renowned rock music star (who was also an enthusiastic environmentalist) took pity on a tank of live lobsters at a seafood restaurant in Geneva. He bought them all for £1,000 and returned them to freedom at the water's edge. The water in question was that of Lake Leman, so the lobsters were doomed to a long and painful death in the fresh water. Some campaign-ers have released hens from batteries, and mink from fur farms, only to see the creatures subjected to new torments by an envi-ronment for which they are unfitted.

In the new millennium, let us set out to relish life in all its majesty. A new understanding of the living world offers tremendous benefits to us all. We can come closer to our children, and understand how important parental input is to every child. We could improve farms and make them more humane, make the veterinarian's office more welcoming to animals, understand the needs of plants and begin to tune in to the majestic cadences of life. A new celebration of life as a whole could help us feed the hungry, cure the sick and mend the wounded.

All life has its language. Our challenge in this new century will be to understand it.

Further Reading

Ackerman, Diane, 1990, *Natural history of the senses*, London: Chapmans.
Addicott, F. T., 1982, *Abscission*, Berkeley and Los Angeles: University of California Press.
Aphalo, P. J., and Ballare, C. L., 1995, On the importance of information-acquiring systems in plant–plant interactions, *Functional Ecology*, 9: 5–14.
Attridge, T. H., 1990, *Light and plant responses*, London: Edward Arnold.
Ballaré, C. L., 1994, Light gaps: sensing the light opportunities in highly dynamic canopy environments, in: *Exploitation of environmental heterogeneity by plants. Ecophysiological processes above and below ground*: 73–110 (Physiological Ecology), San Diego: Academic Press.
Ballaré, C. L., Scopel, A. L., Jordan, E. T., and Vierstra, R. K., 1995, Signaling among neighboring plants and the development of size inequalities in plant populations, *Proceedings of the National Academy of Sciences of the USA*, [in press].
Bayer, Frederick, and Owre, Harding, 1968, *Free-Living Lower Invertebrates*, New York and London: Macmillan.
Bayssade-Dufour, Christiane, 1979, *L'appareil sensoriel des cercaires*, Paris: Editions du Muséum.
Binet, A., 1888, *The psychic life of microorganisms – a study in experimental psychology*, Chicago: Open Court.
Bond, Nigel W. (ed.), 1984, *Animal models in psychopathology*, London: Academic Press.
Bradbury, Jack W., and Vehrencamp, Sandra L., 1998, *Principles of Animal Communication*, Sunderland, MA: Sinauer Associates.

Bruin J., Sabelis, M. W., and Dicke, M., 1995, Do plants tap SOS signals from their infested neighbours?, *Trends in Ecology and Evolution*, 10: 167–70.

Burton, J., 1970, *Animal senses*, Newton Abbot: David & Charles.

Byrne, Richard, 1995, *Thinking Ape – Evolutionary Origins of Intelligence*, Oxford: University Press.

Coleman, James (ed.), c.1990, *Development of sensory systems in mammals*, New York: Wiley Interscience.

Crail, T., 1983, *Apetalk and whalespeak, the quest for interspecies communication*, Chicago: Contemporary Books.

Darwin, Charles, 1875, *The movements and habits of climbing plants*; 1880, *The power of movement in plants* (new edition, London: Pickering, 1988–9).

Downer, John, 1988, *Supersense, perception in the animal world*, London: BBC Publications.

Dugatkin, Lee Alan, and Godin, Jean-Guy J., 1998, How females choose their mates, *Scientific American*, 278 (4): 56–61.

Dusenbery, D., 1992, *Sensory ecology – how organisms acquire and respond to information*, New York: W. H. Freeman.

England, R., Hobbs, G., Bainton, N., and Roberts, D. McL., 2000, *Microbial signalling and communication*, Cambridge: University

Experimental Psychology Society, c.1986, *The use of animals for research by psychologists*, St Andrews: Experimental Psychology Society.

Ford, Brian J., 1975, Microscopic Blind Spots, [leading article for] *Nature*, 258: 469, 11 December.

Ford, Brian J., 1976, *Microbe Power – Tomorrow's Revolution*, London: Macdonald & Jane's; New York: Stein and Day; Tokyo: Kodan Sha.

Ford, Brian J., 1996, Cells, the ultimate microdots, *Biologist*, 43 (2): 96.

Fouts, Roger, 1997, *Next of Kin*, London: Michael Joseph.

Greuet, O., 1968, Organisation ultra-structurale de l'ocelle de deux Péridiniens Warnowiidae, *Erythropsis pavillardi* (Kofoid et Swzy) et *Warnowia pulchra* (Schiller), *Protistologica*, 4: 209–30.

Griffin, D., 1981, *The question of animal awareness*, Los Altos, CA: Kaufman.

Griffin, D., 1981, *Animal thinking*, Cambridge, MA: Harvard University Press.

Hall, Rebecca, 1984, *Voiceless victims*, Hounslow: Wildwood House.

Hanlon, R. T., and Messenger, J., 1996, *Cephalopod behaviour*, Cambridge University Press.

Hart, James, 1990, *Plant tropisms and other growth movements*, London: Unwin Hyman.

Haupt, W., and Feinleib, M. E., 1979, *Physiology of [plant] movements*, Berlin: Springer.

Hidgson, Edward, and Mathewson, Robert (eds), 1978, *Sensory biology of sharks, skates and rays*, Arlington: Department of the Navy.

Hoage, R., and Goldman, L., 1986, *Animal intelligence: insights into the animal mind*, Washington, DC: Smithsonian Press.

Jay, Ricky, 1987, *Learned pigs and fireproof women*, London: Hale.

Katterman, Frank (ed.), c.1990, *Environmental injury to plants*, Washington DC: American Chemical Society.

Keehn, J. D. (ed.), 1979, *Psychopathology in Animals*, London: Academic Press.

Lambton, Lucinda, 1992, *Magnificent menagerie*, London: HarperCollins.

Mattiacci, L., Dicke, M., and Posthunus, M. A., 1995, Beta-glucosidase: elicitor of herbivore-induced plant odors that attract host-searching parasitic wasps, *Proceedings of the National Academy of Sciences of the USA*, 92: 2036–40.

Michell, John, c.1982, *Living wonders, mysteries and curiosities of the animal world*, London: Thames & Hudson.

Muñoz-Cuevas, Arturo, 1981, *Développement, rudimentation et régression de l'oeil chez les Opilions (arachnida)*, Paris: Editions du Muséum.

Noback, Charles (ed.), 1978, *Sensory systems of the primates*, London: Plenum.

Passarin d'Entrèves, P., and Zunino, N., 1976, *The Secret Lives of Insects*, London: Orbis Publishing.

Pearse, V. & J., and Buchsbaum, M. & R., 1987, *Living Invertebrates*, Palto Alto, CA: Blackwell Scientific and Boxtree Press.

Penard, E., 1938, *Les infinitemens petits dans leurs manifestations vitales*, Geneva: George et cie.

Sebeok, T., and Rosenthal, R. (eds), 1981, The Clever Hans Phenomenon, communication with horses, whales, apes and people, *Annals of the New York Academy of Sciences*, 364.

Shuker, Karl, 1991, *Extraordinary animals worldwide*, London: Hale.

Simons, Paul, 1992, *The Action Plant*, Oxford: Blackwell.

Sinnott, E. W., 1950, *Cell and psyche, the biology of purpose*, University of North Carolina Press.

Treshow, M., 1970, *Environment and plant response*, New York: Wiley.

Wade, N., 1980, Does man alone have a language? Apes reply in riddles, and a horse says neigh, *Science*, 208: 1349–51, 20 June.

Index

Numbers in bold denote major section/chapter devoted to subject.